Fig. 1.1b G. K. Chesterton

The wicked Grocer groces

In spirits and in wine,

Not frankly and in fellowship

As men in inns do dine;

But packed with soap and sardines

And carried off by grooms,

For to be snatched by Duchesses

And drunk in dressing-rooms.

G K Chesterston "Song against Grocers"

BOUDOIR LABELS
1753 - 1987

BY JOHN SALTER

The Wine Label Circle, 2012

Published 2012 by

The Wine Label Circle

Hon Secretary email: winelabels@hotmail.co.uk

Website http://www.winelabelcircle.org

ISBN 978-0-9572523-0-1

British Library Cataloguing Data.

A catalogue record for this book is available from the British Library

To order a copy of this book, contact The Wine Label Circle

Designed and edited by Robyn Mercer

Photographs by Robyn Mercer and Stephanie Cripps

Printed by MPG Books Limited

British Library catalogue

Salter, J.R. (John R.), 1932

Boudoir Labels

Includes bibliographical references, list of illustrations
and index

Set in Minion Pro typeface

BOUDOIR LABELS
1753 - 1987

BY PROFESSOR JOHN SALTER
MA (Oxon), CEnv, FCMI, FCIWM, FRGS, FRSA, ACIArb.

President 1986 - 1987, Hon. Member and Hon. Solicitor of
The Wine Label Circle

In celebration of the Diamond Jubilees of

The Wine Label Circle
and
H.M. The Queen

The Wine Label Circle, 2012

Fig. 1.1c
The Wine Label Circle's Presidential Badge designed by Mark Fitzpatrick and produced by CJ Vander Ltd.

PREFACE

INSPIRED PERHAPS by the publication in 1947 of Dr. Norman Penzer's "The Book of the Wine Label", The Wine Label Circle was founded in April 1952, with the original object being the stimulation and encouragement of research into decanter labels. It publishes a Journal and holds meetings of members. A number of miscellaneous labels were identified and 34 of these, including medicinal and toilet labels, were listed by Major Gray in 1958. These are now called "boudoir labels".

I welcome the publication of Professor Salter's research which not only lists 558 labels but also for the first time includes a commentary on each label, and where possible a photograph, and a review of the historical setting in which boudoir labels were used.

The year 2012 brings the Diamond Jubilee of The Circle and of Her Majesty Queen Elizabeth II, as well as the holding in London of the Olympic and Paralympic Games. The 600th Anniversary of its Guildhall, the Bicentenary of Charles Dickens' birth and the 50th Anniversary of the City of London festival are also being celebrated.

It is thus a very fitting time for The Circle to publish "Boudoir Labels", completing John Salter's trilogy of publications, the others being "Wine Labels" and "Sauce Labels".

As well as thanking John, it is important to recognise the importance of the input of our Past President, Tony Hampton who has steered this book project through to its publication, ably assisted by Tim Kent. We are also indebted to Robyn Mercer our photographer, editor and designer.

Terry Gill
President of the Wine Label Circle

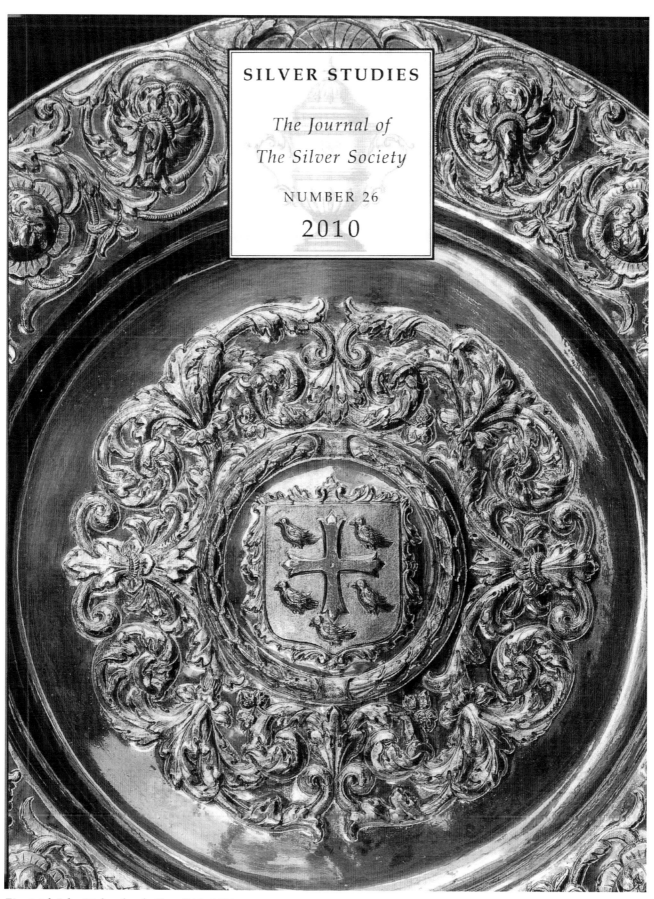

SILVER STUDIES

The Journal of
The Silver Society

NUMBER 26

2010

Fig. 1.1d John Richardson's Alms Dish 1684

FOREWORD

On the significance of Boudoir Labels

Silver is deeply rooted in our social history and has the unique attribute of carrying the record of its origins. For more than a hundred and fifty years, enthusiasts have shared their research into silver, its makers and its functions. Each generation has exploited the rich paper trail left by the precious metals and opened up new questions, or filled in the broad picture created by their predecessors.

The Wine Label Circle has a leading place among the several clubs for specialist interests, pursuing its mission to exchange and disseminate those new discoveries. Honoured for its patronage of research, it has published a journal for many years. Edited and largely written by Professor John Salter, two handsome books on Wine Labels and Sauce Labels, the fruit of collaboration between its members, both broke new ground. They are now joined by this latest achievement. Richly illustrated, with a witty and engaging text, its chapters cover an extraordinary range of material. Sources range from early housekeeping books to apothecaries' lists. These distinctive labels, now categorised for the first time, are helpfully distinguished, not before time, from all those broadly categorised in the past as "wine labels".

As a silver historian, I celebrate this book as a major achievement. It exposes a whole neglected world of boxes, bottles, flasks and their perfumed, delicious, stimulating or even poisonous contents, and sets them in the context of the boudoir, study, cabinet or travelling chariot. In the Diamond Jubilee year, this tribute to a largely neglected female world is a welcome arrival.

Philippa Glanville, FSA
President of the Silver Society and Past-Master of the Company of Arts Scholars

DEDICATION

To Her Majesty The Queen
On the occasion of Her Diamond Jubilee
&
In memory of Her Late Majesty The Queen Mother
Who so much enjoyed visiting auction rooms on viewing days.

REFERENCES & CITATION

The Marshall Collection means The Marshall Collection of Labels for Wines, Spirits, Sauces & Toilet Waters Catalogue and Commentary by M. Strafford and D. Thomas published by The Ashmolean Museum of Oxford in 2005.

The Master List means the alphabetical list of names of wines, spirits, liqueurs and alcoholic cordials on bottle tickets and bin labels as at June 2003 as updated to 30 June 2008 by the First Update and to 30 June 2009 by the Second Update published by the Wine Label Circle.

The Journal is cited as WLCJ preceded by the Volume number and followed by the issue number and the page number.

ACKNOWLEDGEMENTS

The Wine Label Circle is very grateful to The Worshipful Company of Goldsmiths and The Silver Society for their generous financial support for this publication.

The author is most grateful to individual collectors who have contributed information and photographs and especially Robin Butler, Louis Challier, Andrew Collins, Bruce Jones, Roger Pinnington, Gordon Procter, Anthony Taylor, Richard Wells and Rupert Wilkinson; and to the Ashmolean, British, Dental, Ironbridge, Sheffield City, Wellcome (London) and Victoria & Albert Museums, Messrs Bourdon-Smith, Lecurieux and Schredds and the auction houses Bigwoods, Bonhams, Christie's, Sworders and Woolley & Wallis for for all their support and to Robert Sackville-West (August 31, 2010) "Inheritance: The Story of Knowle and the Sackvilles" and Bloomsbury Publishing Plc for the family photograph fig. 1.4 and map fig. 1.3. The assistance given by the Society of Pharmacists of Bruges, the Worshipful Society of Apothecaries of London, Belvoir Fruit-Farms and DR Harris & Co. Ltd. is gratefully acknowledged.

CONTENTS

Fig 1.1

CHAPTER 1

INTRODUCTION

1.1 Silver bottle tickets identified the contents of flint glass decanters from about 1729 onwards. Enamel bottle tickets identified the contents of glass bottles from just before the time of the short-lived Battersea factory (1753-1756). A selection of boudoir labels and boudoir tipples is illustrated in fig.1.1. The earliest known possible boudoir labels made of enamel are HUILE DE VENUS and MUSCAT, a pair of Battersea enamels of similar design by Ravenet (fig. 1.2). The earliest known attributable silver boudoir labels bear the maker's mark of Margaret Binley (1764-1778) for the perfumes EAU DE COLOGNE and MILLE FLUERS and the medicinal AROMATIC VINEGAR, ELDER and VERBENA.

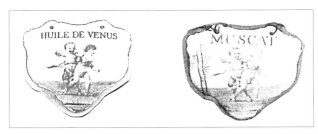

Fig 1.2

1.2 But what is a Boudoir Label? It is easier to define what it is not. It is not normally a wine, spirit, liqueur or alcoholic cordial label dealt with in "Wine Labels" (1.1) and listed in the Alphabetical Master List of Names (1.2) or a sauce, condiment, spice, flavouring, food, herbal, pickle, salad oil, vinegar, ketchup, extract, essence or picnic label dealt with in "Sauce Labels" (1.3) and updated in the 2010 Addendum (1.4). It is in essence any other kind of bottle ticket label suspended usually by a chain. So it would not include, for example, glass bottles with silver tops portraying a ladybird, four leaf clover, white pig and sycamore leaf dated 1905, or with paper labels or with moulded glass titles.

1.3 But why choose "boudoir" as a descriptive title for a suspended bottle label? Wine labels were made to grace glass decanters to identify a wide range of decanted wines drunk in the dining room from about the 1720s. Sauce labels, much smaller, were designed to grace soy frame glass bottles strategically placed on the dining table from about the 1750s, although oil and vinegar frame labels may date from earlier times. Spirit labels were designed for the spirit decanter and the tantalus frame from about the 1780s. None of these were very feminine. Perfume bottles on the dressing room table, toiletry and medicinal bottles on the dressing room shelf, pick-me-up and refreshment bottles in the bedroom, and a range of bottled drinks in the sitting room all required labelling in an appropriate style, reflecting femininity and changing tastes in the household and household management, especially with the onset of art nouveau designs from about 1895, art deco designs from about 1919 and retro designs from about 1950.

1.4 So enamels and porcelains were chosen to create a new look, and to introduce colourful displays with appropriate floral designs. The concept of the boudoir seems to be a fitting description for this style of label. An attempt is made in this book to set boudoir labels into their historical context and wherever possible to illustrate them even where the photography is poor due to the reprographic process.

1.5 Thus the expression "boudoir" where used in this book includes lady's or gentleman's private set of rooms, a sitting room, a drawing room, a withdrawing room (a substantial "withdrawing room" is shown as first floor level on the 1689 Survey Plans of Chevening House in Kent (1.4a) in the triple pile or three rooms wide design), a lady's retiring room (called an ante-room at Powderham Castle in Devon), a bedroom, a dressing room, a powder room, a bathroom, a washroom, a sickroom, a nursery, a packing room, a sewing room and similar accommodation and facilities for servicing such accommodation and its contents. According to Wikipaedia the expression "boudoir" is derived from the French verb "bouder", to pout. Charles Rennie Mackintosh (1868-1928) executed a marvellous stencil design for a "Rose Boudoir" in 1902 which can be seen in the Hunterian Museum and Art Gallery in Glasgow.

1.6   A plan showing the extensive boudoir at Knole in Kent (1.5) running at ground floor level beneath the whole length of the Cartoon Gallery at first floor level is illustrated (fig. 1.3) together with a late Nineteenth Century photograph (1.6) of a corner of the boudoir full of knick-knacks and a confusion of glass decanters, glass bottles and enamelled bottles (fig. 1.4). There was plenty of room for a boudoir grand piano in this type of extensive accommodation!

1.7   Robert Adam's red drawing room at Syon House near London built around 1762 was very

Fig 1.3

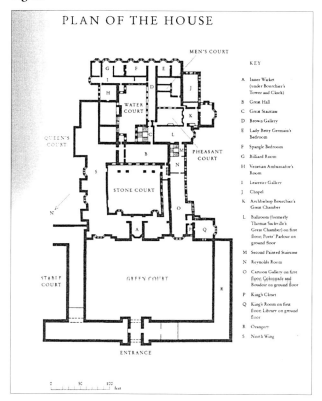

grand. The bedrooms have washing tables set in the alcoves with marble splashbacks. Some marble topped tables or wash-stands stood alone. Some had only two front legs with the back being secured to a wall. Some eighteenth century wooden washstands were triangular shaped designed to fit into the corner of a room.

Fig 1.4

1.8   Warwick Castle's interior house built for the Earls of Warwick has grand accommodation. A report in the Warwickshire Standard for June 1858 of Queen Victoria's short visit on the 16th June states that "the boudoir was a perfect picture fitted up with blue and white satin" wall hangings. It comprises a large room with many gilt framed paintings hung on the walls with comfortable furniture save in one regard. The fifth Earl's memoirs record that Lady Mexborough, an elderly relative, suffered an unfortunate accident with a wooden chair in the boudoir when invited by the Queen during the visit to be seated! The housekeeper in charge, Maria Hume, had obviously failed to spot the defect.

1.9   The point is made in the Official Souvenir Guide to Clarence House that Nash was called to account in 1826 for the fact that the new building was three feet taller than that shown on the approved plans and thus more expensive. The architect replied that he had raised the height of the attic on learning that some of the rooms "were to be appropriated as Sitting Rooms to the Ladies, they being in their original state too low for Apartments so occupied". The ground floor boudoir at Clarence House has undergone many changes. A plan of 1841 shows that certain internal walls had been demolished. Princess Margaret used the larger part of this area as her sitting room or boudoir when she lived at Clarence House before her marriage in 1960. The smaller room and writing room was used by her Lady-in-Waiting. Queen Elizabeth The Queen Mother then converted the area into her so-called Garden Room which functioned as a kind of grand boudoir.

1.10 Belmont House near Faversham in Kent is an example of smaller accommodation. As a private family home it has the benefit of having original furniture and fittings in place, having been in the Harris family ownership since 1801. The first floor level boudoir is located between the South or Blue and the North or Green Bedroom suites (1.7). The butler's bells indicator panel shows under the heading "GREEN" that it activates for the GREEN ROOM, DRESSING ROOM and BOUDOIR. Interestingly the Green (now called North) Bedroom has in its alcove an eighteenth century Vizigapatam ivory veneered five-drawer jewel chest. Standing on top is an oval mirror of similar design surmounted by a small shield bearing the initials of the first Lady Harris. Because of its connection with India one could imagine there being in Lord Harris' withdrawing room (on the ground floor opposite his study/library) a decanter frame with tiger claw labels on the bottles for SHERRY, CLARET and MADEIRA (fig. 1.5) for boudoir tipples.

1.11 The importance of the boudoir is well illustrated in the construction of the neo-Norman Penrhyn Castle in North Wales in the 1830s. The boudoir was situated in

Fig 1.5

prime position on the first floor of the keep, along with Lord Penrhyn's sitting room. The bell board contains two prominent plaques, namely "Ist STry. KEEP LORD PENRHYN's SITTING ROOM" and " Ist STry KEEP LADY PENRHYN's BOUDOIR". On the next floor directly above are the dressing rooms. The Castle, keeping abreast of innovation, also contains a flushing lavatory basin. Before its invention the night table or the library steps were used to hide the chamber pot which had been in use from the Fourteenth Century (1.8). Sometimes it was put under the bed.

1.12 Castle Coole, an Eighteenth Century National Trust property near Ennisikillen in County Fermanagh, Northern Ireland, boasts not only an elegant saloon where guests were received and morning prayers said but also a very fine boudoir or ladies work-room decorated with stately grandeur in the chinoiserie style. Dunster Castle, a National Trust property near Minehead in Somerset, founded in Norman times, was the home of the Luttrell family for more than 600 years. The remodelling in 1868 by Antony Salvin included a splendidly intimate boudoir for Mrs Luttrell. Another example of a boudoir is situated in the late Victorian reproduction chateau built by Baron Ferdinand de Rothschild, Waddesdon Manor in Buckinghamshire, where electricity was installed in 1899. One of the first private houses to be lit by electricity however was Sir David Salomon's House at Southborough in Kent, which even had a sculpture room adjoining the library next to the withdrawing room. The boudoir at Waddesdon can be found on the first floor of the West Wing. It was built in time for the visit of Queen Victoria on the 14th May 1890. Known as the Green Boudoir, it is heavily mirrored with somewhat startling effects! It was in the Green Boudoir that Her Majesty was presented with and gratefully accepted the gift of a jewelled fan. The Grey Drawing Room was also used as a ladies retiring room after dinner.

1.13 An example of an art deco boudoir is that created in 1925 for Lady Baillie by her designer, Armand-Albert Rateau, in the Gloriette of Leeds castle in Kent. Her rooms were on the

Upper Floor. She had a bedroom, dressing room, bathroom and then a large boudoir with painted panelling and theatrical mirrors.

1.14 Boudoir labels other than boudoir tipples can conveniently be grouped into five sub-groups - alphabetically these are medicinal, perfumery, soft drinks, toiletry and writing inks. Some labels belong to more than one sub-group. The names on some boudoir labels are also names of wines, spirits and sauces, which can lead to confusion. So for purposes of easy reference Chapter Two contains an alphabetical list of all boudoir labels covered in this book (there will be updates required as it is not an exhaustive list). Separate sub-lists of medicinal, perfume, soft drink, toiletry and writing ink labels are set out in Chapters Four to Eight. Also are listed boudoir tipples, although these are included in the Master List (1.2), being alcoholic drinks enjoyed in the boudoir such as home made wines, alcoholic cordials, liqueurs, spirits and home produced drinks from the still room (1.9) where distillation took place. These can often be identified by labels (i) not displaying grapes but berries and not displaying vine leaves but floral foliage and (ii) usually being of small size. So Reily and

Fig 1.6

Storer in 1838 made MADEIRA and PORT in the usual wine label size and in 1839 DARK SHERRY and LIGHT SHERRY in boudoir label size for the smaller decanter as illustrated (fig. 1.6). Boudoir tipple labels, often being small in size, made excellent gifts in recognition of hospitality afforded. After all, as was inscribed on a ladies porcelain hand-warmer made around 1740:

> "The Gift is Small,
> Goodwill is All."

1.15 Thus the alphabetical list contains titles of labels (i) which may in fact refer to alcoholic drinks although not of the usual wine label size and design; thus, for example, the list contains a reference to FINE NAPOLEON because it matches the floral decoration of

PEROXIDE; (ii) which may double up both as medicinal labels and as wine and spirit labels; and (iii) which may in fact have been made as a sauce label. Selection has to be based on considerations such as size, the purpose of other labels in the same set, decoration, feel and design, evidence from cellar books and materials used. Small initial labels create problems of attribution and so cannot be regarded as evidential. The titles fall into various categories, depending on whether the user was dressing up, or taking a medicine, a pick-me-up or a cure for headaches, or coping with hot weather or surroundings or a tedious journey, or enjoying a picnic, or getting up in the morning or going to bed at night. Thus cordial labels are found in travelling cases and labels for various kinds of water in picnic chests. That picnics were popular in the Eighteenth Century is shown by the large number of ladies' fans which incorporated picnics into their designs in a central position.

1.16 Size is a good guide, but only a guide, to a label's purpose. The smallest size, a width not exceeding 2cm., is used for initial labels. But they create problems of attribution. Only H for HUILE or HUNGARY water and LW for LAVENDER WATER are included in the list. This is on the basis of the style of engraving which suggests a pairing. Given that the Marshall Collection in the Ashmolean Museum at Oxford of silver labels for wines, spirits, sauces and toilet waters is a reasonably representative collection and taking note of the measurements recorded in Strafford and Thomas' Catalogue (1.10) with its helpful commentary, it appears that single initial labels form the smallest size group with a starting size of around 1.2cm by 1.4cm. But how does one make attributions? Hence:

A could stand for Apple, Anis or Apricot
B could stand for Burgess, Balm, Brandy or Benedictine
C could stand for Chili, Citron or Chartreuse
D could stand for Devonshire, Dactillus or Damson
F could stand for Fish, French or Florence
G could stand for Garlick, Grapefruit or Ginger Brandy
H could stand for Harvey, Huile or Hollands

K could stand for Kyan, Kirchen Wasser or
　　Kirschwasser
M could stand for Mushroom, Methylated or
　　Maraschino
P could stand for Port Sauce, Perfume or Perry
R could stand for Reading, Rose or Ratafia
S could stand for Soy, Seltzer or Soda Water
V could stand for Vinegar, Violette or Vichy
W could stand for Walnut, Woodbine or Water
Y could stand for Yorkshire, Ylang Ylang or
　　Yellow Chartreuse.

1.17　The next size up is the not exceeding 3cm. class. The delightful little silver and silver-gilt perfume, medicinal and toiletry labels rarely exceed 3 cm. in width. Only on rare occasions does one find wine or spirit labels in the 2cm. to 3cm. in width class. Two spirit labels, one for MALT (Whisky) and another for ORANGE GIN, have been noted in the whole of the Marshall Collection as breaking this rule, coming in at 2.6 and 2.3 cm. respectively and presumably intended for use on a boudoir spirit decanter frame. Thus, for example, Thomas Streetin's pair of boudoir labels for TOOTHMIX and EAU DE COL measure 2.1 x 1.1 cm., an unmarked RED VINEGAR measures 2.1 x 1.1 cm.,an unmarked OIL measures 2.2 x 1.2 cm., John Reily's EAU DE COLOGNE, ESSENCE OF ROSE and LAVENDER measure 2.4 x 1.3 cm., RATAFIA measures 2.4 x 1.1 cm., CAMPHOR JULEP measures 2.5 x 1.3 cm., TOOTH MIXTURE, SPIRIT LAVENDER, POISON LAUDANUM, ESSENCE OF GINGER, CAMPHtd SPIRIT and EAU D'ALIBURGH all measure 2.6 x 1.4 cm., ESS GINGER measures 2.6 x 1.4 cm., CAMPHORATED SPIRITS OF WINE measures 2.6 x 1.4 cm., John Reily's pair for JESSAMINE and PORTUGAL measure 2.6 x 1.5 cm., an unmarked LEMON measures 2.6 x 1.3 cm., an unmarked VINEGAR measures 2.7 x 1.3 cm., Reily and Storer's NOYAU measures 2.9 x 1.9 cm., an unmarked ROSE.WATER measures 2.9 x 1.9 cm., a set of four plated fruit juices RAISIN, CHERRY, CURRANT and ORANGE all measure 3.0 x 1.9 cm., Eau de Nil measures 3.0 x 1.5 cm., ESSENCE GINGER measures 3.0 x 1.6 cm., and a cherub's head design EAU DE ROSE measures 3.0 x 2.6 cm. as do the plated BITTERS and R. VINEGAR.

Even the enamel labels follow suit. The small DACTILLUS and WHITE ROSE each measure 3.0 x 2.6 cm.

1.18　The 2-3 cm. class also covers most sauce labels with some exceptions including expansive Regency style decoratives. Thus for example all the sauce labels in the Dr. Richard Wells Collection fall into this class according to the sale catalogue (1.11) except Rawlings and Summers SOY and TARRAGON of 1834 and CAYENNE of c. 1829 (all 3.4 cm.), Boulton's ANCHOVY, CAYENNE and KETCHUP of 1826 (3.2 cm.) and Reily and Storer's ANCHOVY and CHILI of 1829 (3.4 cm.). All the Marshall Collection sauce labels comply except for LEM.PICK and PEPPER of around 1790 and all the boudoir labels in the Wells collection comply.

1.19　The next size up, labels in the 3-4 cm. in width class, is that favoured for spirits, liqueurs and alcoholic cordials drunk in the boudoir and classified as boudoir tipples. Examples in silver are James Hyde's SHRUB of 1799, Parsons, Bennett and Goss's Exeter disc of around 1800 for HOLLAND, John Emes' GIN of 1805, Elizabeth Morley's RUM, BRANDY and GIN OF 1805, Thomas Watson's CHERRY BY (Brandy) and RASBERRY of 1806, Story and Elliott's shell SHRUB of 1810, a BRANDY of 1815, Mary Huntingdon's COGNAC BRANDY of 1822-7, William Knight's BOUNCE and MARASc (Maraschino) of 1828, Isaac Parkin's BRANDY of 1829, Barnards' spirit decanter shaped labels of 1850 for G, W and B, CARLOWITZ of 1880, BENEDICTINE, GINGER BRANDY, CHERRY BRANDY and MARACHINO of 1895, TRAPPISTINE of 1897, ORANGE BITTERS and ORANGE CURACAO of 1916 and MINTE and COINTREAU of 1922. However one possible contender for recognition as a boudoir label in this class is Robert Gray's CORDIAL which measures 3.5 x 1.9 cm. and could be non-alcoholic. Another is INDIAN TEA but this is a porcelain label with different manufacturing criteria. It has a width of 3.8 cm.

1.20　The next size up, labels in the 4-5 cm. in width class, is that favoured by wine and spirit labels. Large decanters, such as those containing water, need larger labels. So Joseph Willmore's HUNGARY WATER measures 4.3 x 2.4 cm., Bent & Tagg's DENTIFRICE WATER measures 4.7 x 2.5 cm. and the plated MALVERN WATER measures 4.9 x 1.0 cm. Boudoir labels falling within this group include ALBAFLOR (Old Sheffield Plate), B,CURRANT (Electroplate), C. De ROSE, CHERRY, CURRANT (silver 1804 and 1812), EAU DE LUBIN (enamel), EAU DE TOILETTE (enamel), GINGER-CORDIAL (plated), GOOSEBERRY, GRAPE, MILK PUNCH, MINT (plated), NOYAU (silver 1827), NOYEAU (silver 1820), ORANGE (silver 1820), PARSNIP (silver c.1785), PEACH (plated), PEPPERMINT (silver 1844), POISON (unmarked silver), RAISIN (silver 1820), RASBERRY (plated), RATAFIA (plated), TONIC (enamel), VINEGAR (unmarked silver) and W.CURRANT. BURGUNDY in a small decanter with a small label such as that made by William Key of London in 1784-5 4.7cm. in width might have been drunk in the boudoir. However a label for HARVEY has been noted of 4.25cm in width which one expects to refer to the sauce but might refer to a Bristol Cream Sherry!

1.21 The next size up, labels in the 5-6 cm. in width class, is indicative of heavy use in significant quantities. Boudoir labels falling within this group include enamels such as those for ANISETTE, CRÈME DE VANILLE, DENTIFRICE, EAU BORIQUEE, EAU DE TOILETTE, ELIXIR, VERVEINE, VIOLET, WHITE LILAC and WHITE ROSE, silver such as those for ALBAFLORA (Binley), HUILLE DE ROYAUX (Halford), KIRSCHWASSER (Reily and Storer) now abbreviated to Kirsch, and plated such as CLOVES, CORDIAL, CRÈME DE NOYEAU, PEPPERMINT, MILK, L. JUICE, FLORIDA WATER, RAISEN and RASPBERRY.

1.22 The largest size, labels in excess of 6 cm in width class, caters for some interesting potential boudoir labels including ANNISCETTE (plated), CHIPRE (enamel), CHYPRE (porcelain), LA ROSE (silver), MUSCAT (enamel) and SCYRUP (plated). ANNISCETTE and SCYRUP not only have vine and grape borders but also are part of a set with WHKSE. They are therefore probably not boudoir labels. Size throws doubt upon the candidacy of the others.

1.23 Illustrated in fig. 1.7 for the purpose of size comparison are 77 labels arranged in 11 rows A to K of 7 labels each. Potential boudoir labels are located at C5 (LACHRYMA), E2 (RATAFIA), E3 (PARFAITE AMOUR), H2 (GRAPE), J7 (MALVERN WATER), K5 (NOYEAU) and K6 (MILK PUNCH). Size comparisons are by no means conclusive.

Fig 1.7

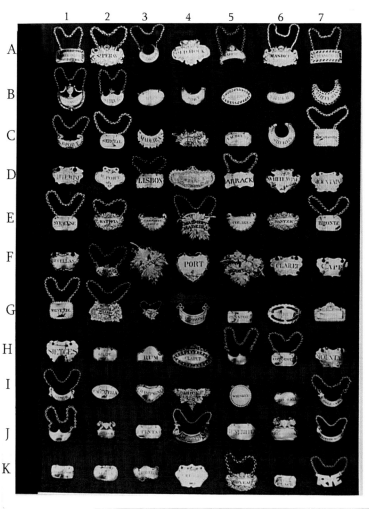

Fig. 1.8 shows 18 labels arranged in 6 rows of 3 labels each. Potential boudoir labels are located at A3 (CINNAMON), C3 (TANGERINE), D1 (WHITE CONSTANTIA), D3 (RED CONSTANTIA), E1 (Crème De Noyeau) and E3 (Crème De Thé). The two Cremes are very small labels. Finally fig. 1.9 from a Bonham's sale of fine silver and vertu catalogue shows the tiny VIOLETTE and EAU DE PORTUGAL labels (Lot 505) which undoubtedly fall within the boudoir category whilst WHISKY (Lot 504) and Lisbon (Lot 501) might well have been drunk in the boudoir being of such small size.

1.24 Searching cellar books for mention of any boudoir title is an interesting test, especially when provenance has been established. For example, the Burghley House cellar book for April 1799 contained no mention of any title set out in the Alphabetical List of Boudoir

Fig 1.8

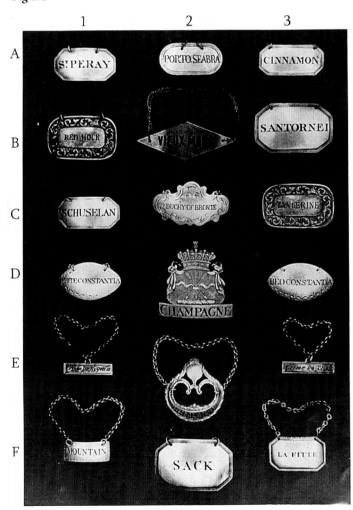

Titles contained in Chapter Two. Wine sales from big country house cellars are also a useful reference point. For example, Christie's sold the contents of a "Cellar of Choice Old Wines of the Rt. Hon. Lord Ashburton, Deceas'd, removed from Buckenham, Norfolk" which contained no mention of any title mentioned in the Alphabetical List.

1.25 How were the labels displayed in the boudoir? In 1893 Baroness Staffe wrote her etiquette and advice manuals about Victorian London. In "The Lady's Dressing Room" she set out her advice on furnishings and indispensible accessories. "The dressing-room of every well-bred woman should be both elegant and comfortable in proportion to her fortune and position; it may be simply comfortable if its owner cannot make it luxurious, but it must be provided with everything necessary for a careful toilette". The dressing–room was where she received her friends. "There must be two tables, opposite to each other, of different dimensions, but the same shape. The larger table is meant for minor ablutions, and on it should be placed a jug and a basin, which should be chosen with taste and care". A Mason's Patent Ironstone China set, pattern 3961, comprising water jug, basin, soap dish, toothbrush holder and two chamber pots, is illustrated (fig. 7.1) made by GM and CJ Mason at Fenton Works in Staffordshire between 1813, the year of the patent, and 1829, or the Minton set at Belmont marked with the initials of Lord Harris or a George Jones set which included a ewer, with an attached carrying handle marked "1st quality", pattern number 15793, made at Crescent Pottery in Stoke-on-Trent between 1907 and 1911, would have fulfilled these requirements. "The table is draped to match the walls; above it should run a shelf, on which are placed the bottles for toilet waters and vinegars, dentifrices and perfumes, the toilet bottle and glass, etc" At Belmont a curved marble topped table was fitted into the alcove. The other table had a mirror meant for the operation of hair dressing. Bottles of perfume and of scented oil or pomades "which should be chosen with artistic taste" had their own place on this table, along with a jewel box (fig. 1.10). The Baroness goes on to describe the arrangements and appointments for the bath-room if the house was lucky enough to have one. "All the articles one requires when bathing" are put on "small shelves of marble". Jewels and trinkets should be kept in the bed-room. "Beside the actual baths, there should be in the bath-room a couch or ottoman, whereon to repose after the bath," and "a little table, in case one would wish to have a cup of tea". A tea caddy therefore was an essential boudoir item.

1.26   What a contrast to the sheer elegance of the third quarter of the eighteenth century as demonstrated by the Shrewsbury architect Thomas Farrolls Pritchard of Ironbridge fame in his interior design of the Blue Room boudoir in Croft Castle, a castellated manor house at near Leominster in Herefordshire. He used trompe d'oeil gilt bosses on blue painted panelling to create a unique atmosphere. The

lady of the house, Lady Croft, had her portrait in oils specially set in an elaborate fireplace as a focal point.

1.27   In summary it is not possible to attribute labels to the boudoir category with any certainty in many cases. Accuracy may well be assisted when labels come in groups with the same design, maker or date but with a variety of titles, to adorn for example a soy frame, a small spirit decanter, a boxed perfume set, a medicine chest, a dressing case or a picnic basket. Distinguishing soy labels from boudoir labels may not always be easy. Toiletry labels were generally displayed on square bottles whereas sauce labels adorned round bottles. Early sauce labels were curved to fit the shape of the bottles. The author hopes that this little book may assist in classification and understanding and develop debate and discussion. A stanza (1.12) in G.K. Chesterton's "Song against Grocers" perhaps sums up the position (see fig. 1.1b). Georgiana, Duchess of Devonshire, certainly drunk a lot.

Fig 1.9

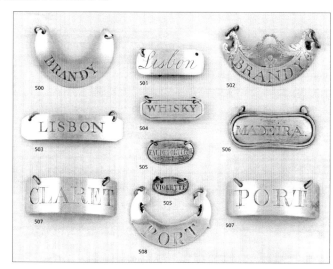

NOTES

1.1   Wine Labels 1730-2003, a worldwide history, edited and co-authored by John Salter, Antique Collectors' Club with The Wine Label Circle, 2004.

1.2   Alphabetical Master List of Names of Wines, Spirits, Liqueurs and Alcoholic Cordials on Bottle Tickets and Bin Labels as at 31 January 2010, The Wine label Circle, 2010.

1.3   Sauce Labels 1750-1950 by John Salter, Antique Collectors' Club with The Wine Label Circle, 2002.

1.4   Sauce Labels 2010 Addendum by John Salter, The Wine Label Circle, 2010.

1.4a   See "Chevening House-Seat of British diplomacy", John Goodall, Country Life 29.2.2012, pp 52-59.

1.5   See "Inheritance – The Story of Knole and the Sackvilles", Robert Sackville-West, Bloomsbury, 2010.

1.6   Ibid., insert between pages 138 and 139.

1.7   Belmont Guidebook, pp18-19

1.8   See further C.C.Stevens, "Privies of Perfection", Antique Collecting, April 2011.

1.9   Still rooms are preserved in a number of stately homes. Whilst the cellar was under the control of the Butler, still rooms were under the control of the Housekeeper working with the Cook. (Fig. 1.1a).

1.10   The Marshall Collection of Labels for Wines, Spirits, Sauces and Toilet Waters Catalogue and Commentary by Molly Strafford and Duncan Thomas, The Ashmolean Museum, Oxford, with The Wine Label Circle, 2005.

1.11   Woolley and Wallis (Woolleys) sale catalogue, 25.1.2011, Lots 520-602.

1.12   Brought to the author's attention by Mr Timothy Kent.

Fig 1.10

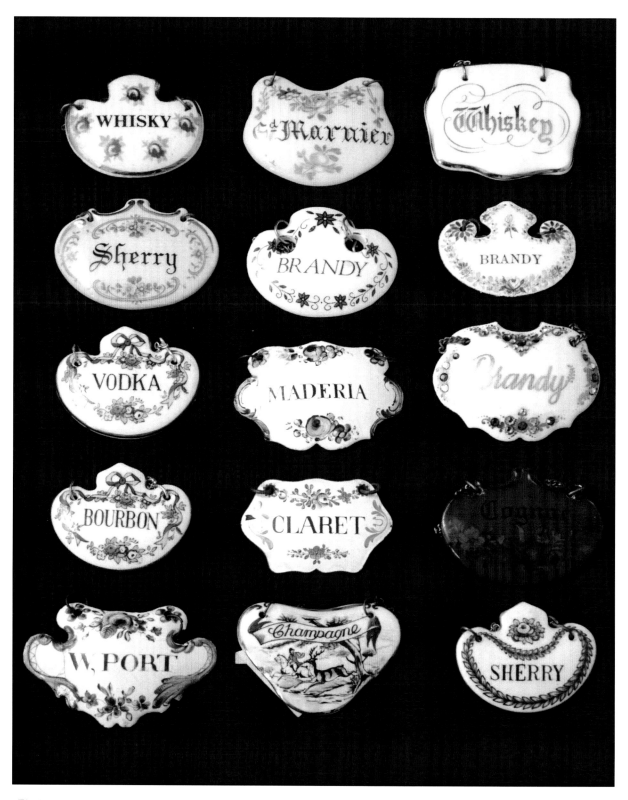

Fig 2.1

CHAPTER 2

TITLES

2.1   This Alphabetical List of selected boudoir labels contains variant spellings or presentations as observed on labels. Selection is based upon the principles set out in Chapter One (Introduction). The List includes labels from all five sub-groups – medicinal, perfumery, soft drinks, toiletries and writing inks, but does not necessarily include all boudoir tipples for which see Chapter Twelve. Fig. 2.1 shows a selection of tipples. For a detailed commentary on the 571 titles see Chapter Nine. For an historical introduction and list of sub-group labels see Chapter Four on medicines, Chapter Five on perfumes, Chapter Six on soft drinks, Chapter Seven on toiletries and Chapter Eight on writing inks.

1a.	A BRANDY	34.	ATTAR OF ROSES	66.	BOUQUET DE ROI
1.	ACID	35.	AURQUEBUSADE	67.	BOUQUET DU ROI
2.	Acqua di Parma	36.	B	68.	BRANDY
3.	ACQUA D'ORO	37.	BALM	69.	BRISTOL WATER
4.	ALBA	38.	BARLEY WATER	69a	C BRANDY
5.	ALBAFLOR	39.	BATH ESSENCE	69b.	C,BRANDY
6.	ALBA FLORA	40.	BATH POWDER	70.	C. PORT
7.	ALBA-FLORA	41.	BATH SALTS	71.	C. De ROSE
8.	ALCOHOL	42.	B,CURRANT	72.	C. DE ROSE
9.	ALKOHOL	43.	B.CURRANT	73.	CACAO
10.	ALMOND CREAM	44.	Benedict	74.	CALAMITY WATER
11.	AMEIXORIA	45.	BENZIN	75.	CAMPH DROPS
12.	AMHONIA	46.	BENZOIN	76.	CAMPHtd SPIRIT
13.	AMMONIA	47.	BENZOIN LOTION	77.	CAMPHOR JUICE
14.	AMONIA	48.	BENZOIN SALTS	78.	CAMPHOR JULEP
15.	ANGELICA	49.	BENZOIN SOLUTION	79.	CAMPHORATED
16.	ANIS	50.	BERGAMOT		SPIRITS OF WINE
17.	ANISEED	51.	BICARBONATE OF	80.	CAPILARE
18.	ANISETTE		SODA	81.	CAPILLAIRE
19.	Anisette blanc	52.	BICARB SODA	82.	CAPISCUM
20.	ANNESEED	53.	BICARD DE SOUDE	83.	Carnation
21.	ANNICETTE	53a.	BITTER	84.	CASTOR OIL
22.	ANNISCETTE	54.	BITTERS	85.	CELLADON
23.	ANNISEED	55.	BLACK	86.	CHARTREUSE
24.	ANNISETTE	56.	BLACKBERRY	87.	CHERRY
25.	APPLE	57.	BLACKCURRANT	88.	CHIPRE
26.	APRICOT	58.	BLACKCURRANTS	89.	Chloroform
27.	AQUA MIRABEL	59.	BORACIC	90.	CHYPRE
28.	ARBAFLOR	60.	BORACIC ACID	91.	CINNAMON
29.	ARISETTE	61.	BORACIC LOTION	91a.	CINQ A SEPT
30.	AROMATIC VINEGAR	62.	BORACIC WATER	92.	CITRATE MAGNESIA
31.	ARQUEBUSADE	63.	BORIC	93.	CITROEN
32.	ARQUEBUZADE	64.	BORIC ACID	94.	CITRON
33.	ASTRINGENT	65.	BOUQUET	95.	CITRONELLE

TITLES

394.	NOYEAU	441.	RAISIN	492.	SPIRIT LAVENDER
395.	NOYEO	442.	RASBERRY	493.	SPRUCE
396.	NOYEU	443.	RASIN	494.	Star Mise
396a.	ODC	444.	RASPAIL	495.	STRAWBERRY
396b.	O.D.C	445.	RASPBERRY	496.	SULTANA
396c.	ODV	446.	RASPBERRY VINEGAR	497.	SURFEIT WATER
396d.	O.D.V	447.	RATAFIA	498.	SWEET
396e.	ODV de Danzic	448.	RATAFIA DE	498a.	Sweet
397.	OEILLET		FLORENCE	499.	Sweet Pea
398.	OIL	449.	RATAFIER DE FLEUR	500.	TANGERINE
399.	OPPOPONAX		D'ORANGE	501.	TAR WATER
400.	ORANGE	450.	RATAFIAT	502.	THISTLE JUICE
401.	ORANGE BITTERS	451.	Ratafiat de fleur d'orange	503.	TISSANNE
402.	ORANGE CURACAO	452.	RATAFIE	504.	TOAST
403.	ORANGE FLOWER	453.	RATIFEE	505.	TOILET LOTION
404.	Orangeade	454.	RATAFIER	506.	TOILET POWDER
405.	Orgeat	455.	RATIFIA	507.	TOILET VINEGAR
406.	PAFAIT AMOUR	456.	RATIFIE	508.	TOILET WATER
406a.	Paniagua	457.	RATTAFIER	509.	TONIC
407.	PARFAIL AMOUR	458.	RATTFIER	510.	TOOTH MIX:
408.	PARFAIT AMOUR	459.	R. CONSTANTIA	511.	TOOTH MIX 1829
409.	PARFAIT D'AMOUR	460.	RED	512.	TOOTHMIX
410.	PARFAITE AMOUR	461.	RED CONSTANTIA	513.	TOOTH MIXTURE
411.	PARSLEY	462.	RED CURRANT	514.	TWININGS BLENDING
412.	PARSNIP	463.	RED VINEGAR		ROOM 1706
413.	PEACH	464.	RONDELETIA	515.	VANILLE
414.	P. MINT	465.	ROSA	516.	VERBENA
415.	PEPPERMINT	466.	ROSE	517.	VERVEINE
416.	PERFUME	467.	Rose	518.	VERVIENE
417.	PEROXIDE	468.	Rosebud Perfume	519.	VICHY
418.	PICK-ME-UP	469.	ROSE GERANIUM	520.	VIELLE FINE
419.	PLUM	470.	ROSEMARY	521.	VIELLEFINE
420.	POISON	471.	ROSE WATER	522.	VINAIGRE
421.	POISON LAUDANUM	472.	ROSE.WATER	523.	Vinaigre Ordinaire
422.	POMADE	473.	ROSE-WATER	524.	VINEGAR
423.	POMAR	474.	ROSEWATER	525.	VIOLET
424.	POMARD	475.	ROSE WATER &	526.	VIOLET DE PARME
425.	POMERANS		GLYCERINE	527.	VIOLETE
426.	POMERANZ	476.	Rp BERRY	528.	VIOLETTE
427.	PONDS EXTRACT	477.	SAL VOLATILE	529.	VIOLETTE DE PARME
428.	PORTUGAL	478.	SAVRON	530.	VIOLLETE
429.	POTASH	479.	S. WATER	531.	WATER
430.	POUSSE L'AMOUR	480.	SCUBAC	532.	WATER SOFTNER
431.	Pt George	481.	SCYRUP	533.	W. CONSTANTIA
432.	PRUNES	482.	SELTER	533a.	WHITE CONSTANTIA
433.	P. VINEGAR	483.	SELTERS	534.	W. CURRANT
434.	QUESTCHE	484.	SELTZER W.	535.	WHITE CURRANT
435.	QUETSCH	485.	SELTZER WATER	536.	WHITE LILAC
436.	QUETSCHE	486.	SEVE	537.	White Lilac
436a.	QUETSH	487.	SHERBERT	538.	WHITE ROSE
437.	QUUETCHE	488.	SHERBET	539.	WHITE VINEGAR
438.	R.CURRANT	489.	SODA	540.	WOODBINE
439.	R. VINEGAR	490.	SODA WATER	541.	YLANG-YLANG
440.	RAISEN	491.	SOURING		

FRONTISPIECE.

TIS EDUCATION, HEAVENLY ART!
DOES EVERY NEEDFUL AID IMPART,
IT GIVES THE GRACES ALL TO SHINE,
AND MAKES THE HUMAN FORM DIVINE.

London, Published by Tho.̅ Kelly, 17, Paternoster Row, March 20, 1824.

Fig 2.2 Frontispiece to "The Female Instructor or Young Woman's Companion and Guide to Domestic Happiness", Thomas Kelly, 1824.

Fig 3.1

CHAPTER 3

TRAVELLING BOXES

3.1 From the times of George III until Edwardian times the wooden travel box or chest had become an essential item of luggage before giving way to the leather case or bag. In the five back row sections of the box or chest the far left and the far right were taken up with two or four square, often cut, glass bottles with usually silver or silver-gilt caps which were adorned with suspended silver or silver-gilt labels describing the contents. A fine example of a Victorian travelling box is illustrated at fig. 3.1. The contents were removed from the box and displayed during the visit on a table in the boudoir.

3.2 The earliest known of these silver travelling box labels bear the mark of Margaret Binley, 1764 -1778 and were for the perfumes EAU DE COLOGNE and MILLE FLUERS and the medicines AROMATIC VINEGAR and VERBENA. In 1815 or thereabouts some oblong shape silver-gilt pairs of labels were made. One pair was for the medicinal CAMPHOR JULEP, named after camiphora, a fragrant plant of the mint family and julep from the Arabic gulab, which is a sweet drink made from sugar syrup to help the medicine go down and for the perfume ROSE WATER. Another pair was for EAU DE COLOGNE and ROSE WATER (both perfumes). Likewise around this time John Reily made a silver-gilt shell and scroll label for LAVENDER (a perfume). In London in 1820 Charles Rawlings made an oblong silver-gilt beaded edge label for ROSE WATER and a shell and scroll silver-gilt label for EAU.DE.COLOGNE. A little later on in Elgin William Ferguson made (but not in silver-gilt) a tiny reeded edge but otherwise plain VIOLETTE and a slightly larger oblong reeded edge but otherwise plain EAU.DE.PORTUGAL. Yet more examples were made a little later on by Rawlings and

Summers. In 1836 they made a pair of shaped labels with scalloped edges for LAVENDER WATER and ROSE WATER. In 1837 they made a pair of reeded octagonals in silver-gilt for PARFAIT D'AMOUR and FLEUR D'ORANGE, perfumes signifying perfect love.

3.3 But what about the boxes or chests, sometimes called "necessaires de voyage" and later on "vanity cases"? A Sixteenth Century example is illustrated in fig. 5.1. It was in use in Germany around 1589. Some of the silverwork on objects relating to eating and drinking was made by Nicolaus Schmidt. Two unmarked silver-gilt capped tall and thin scent bottles were included, fitting snugly into two specially designed spaces in the box, along with a scent flask (3.1) of the usual oval shape of those times. It is unlikely that they carried labels because devices on the cap provided the key to differentiation between the two opaque glass bottles. An early octagonal reinforced metal campaign medical chest, made around 1590 to 1610, with carrying handles, has been noted. It contained 10 glass stoppered bottles, with hand-written adhesive paper titles. Many of these related to herbal remedies. One bottle still contained traces of senapod, a well known laxative. Dr. William Turner set the scene with his three-volumed "A New Herbal" of 1551. He followed this up with a treatise on baths showing medical benefit and then in 1568 published his "A New Book on the Natures and Properties of all Wines".

3.4 John Aubrey, one of the founders of the Royal Society established by Royal Charter in 1662, was a man of varied interests who encouraged the publication of scientific books, including Robert Morison's volumes on botany. His colleague John Wilkins published in 1668 his thoughts about the creation of a new language in which he was already corresponding with others called "An Essay towards a Real Character and a Philosophical Language". It is against this background that another publication in the same year should be evaluated. It was an herbiary by Aubrey's friend Andrew Pascall (1631-1696), a clergyman from Somerset. It contained "Botanick Tables" designed to enable horticulturalists to learn about and indeed memorise the new language. Significantly it was his second Botanick Table listing herbs considered according to their flowers (3.2) which gives a clear indication of the state of knowledge about herbs in the later seventeenth century.

3.5 Particulars of an eighteenth century travelling case have been given in the Journal (3.3). Thought to have been of Dutch origin and of around 1740 in date, it was made of oak clamped at the corners with metal clamps, 35 cm. square and 30 cm. high, with two carrying handles. It was not an apothecary's box because the bottle titles were not in abbreviated Latin. The titles were in English. It could have been a campaign box. The titles were CLOVE WATER, COLLICK WATER, NUTMEG WATER, ORANGE WATER, CELERY WATER, PEACH WATER, GOULD WATER, ANNESEED, CUMMIN WATER, LIMOEN WATER, AQUA MIRABEL, BARBADOS CITROEN and CINNAMON. The titles , being on stick-on paper labels, do not appear on any List, other than in the form of suspension labels for ANISEED, CLOVE WATER, GULDEWATER, CITROEN and CINNAMON.

3.6 An early example of a travelling box made around 1750 and lined with cherry velvet contained no scent bottles even though its cartouche was inscribed "Souvenir de Annette Alpruni, nee Contesse de Lutti". The Counts de Lutti came from the Tyrol and were Knights of the Holy Roman Empire. The contents of the box were of masculine appeal. They may not have been the original items, albeit the silver is contemporary (3.4).

3.7 Edward, Duke of Kent and of Strathearn

(fig. 7.6), the fourth son of King George III and father of Queen Victoria, had a travelling box with silver fitments made for him by John Holloway in London in 1788 (fig. 7.7). It had seven glass bottles, with unmarked silver mounts, in three sizes, presumably to assist in identifying contents as there would appear to have been no labels. The box contained toiletry items, shaving kit and writing materials. The silver toothbrush holder had six detachable toothbrushes (3.5).

3.8 Pierre Leplain, of 36 Rue Saint-Eloi in Paris, made the silver for a travelling box of around 1805 (fig. 7.8). As well as the usual razors and toilet items it contained a large oval silver dish, a teapot, a cup and saucer, the seal of the Marquis de Caulaincourt and four silver topped cut glass bottles of three sizes, one of which appeared to have a suspended silver label attached to it (3.6). Between 1809 and 1819 Pierre Leplain made silver items for a large box containing dining equipment (including candlesticks made by Marc Jacquart) in Paris and eleven silver mounted cut-glass perfume bottles which certainly would have required indentifying. They were presumably at some time distinguished by labelling. The box also contained 47 toiletry articles of various kinds. It was made by Hebert, of 20 Palais Royal in Paris, who was a noted maker of elegant boxes (3.7).

Fig 3.2

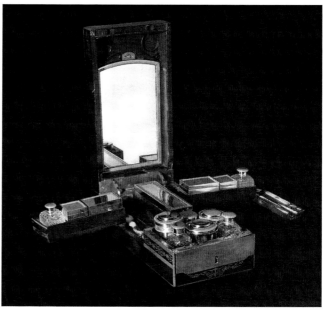

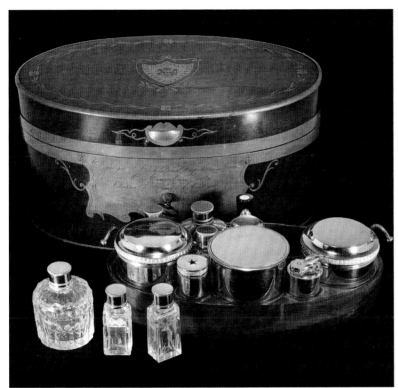

Fig 3.3

Fig 3.4

3.9 Pierre-Noel Blaquiere of Paris during the period 1809 to1819 included in his necessaire de voyage (illustrated in fig. 3.2) four square cut glass bottles with silver gilt covers in two sizes (3.8) and during the period 1819 to1838 he made a compact necessaire having four facetted glass bottles with unmarked silver covers (3.9). Martin-Guillaume Biennais of the Singe Violet in the Rue St. Honore in Paris, who was appointed the Imperial Goldsmith in 1804, included in his necessaire de voyage (fig. 3.3) of about the same period 1809-1819 six cut glass bottles with silver gilt covers, in three different sizes. He was a specialist maker, supplying campaign travelling boxes to officers in the Egyptian and Italian campaigns, including it is said Napoleon himself. Prince Louis Napoleon presented a box to Armand Laity, one of his great supporters, in 1843 (3.10). Maire, another Imperial Box-maker, of 154 Rue St. Honore, employed at this time Denis-Francois Franckson to make most of the silver for a substantial travelling box possibly for Marie-Julie Clary Bonaparte (the initials engraved on the plaque are M.C.B.), who was the Emperor's sister-in-law and at this time (1808-1813) Queen of Spain (3.11). This had

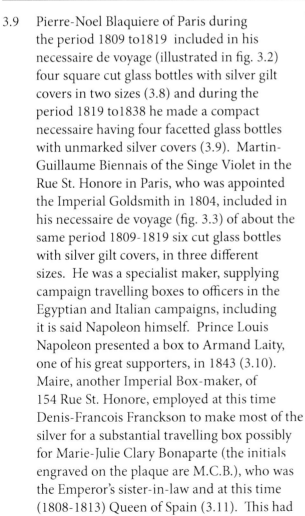

eleven rather large cut glass bottles with silver gilt stoppers, twenty-three toilet articles and a variety of other boudoir necessities (see fig. 3.4).

3.10 In England in 1825 D. Edwards, holder of a Royal Warrant, of 21 King Street off

Fig 3.5

Fig 3.6

Bloomsbury Square, made an Officer's brass
bound rosewood and walnut campaign chest
(fig. 3.5) identified by the owner's initials
(C.C.A.), which included two glass toilet bottles
mounted by Allen Dominy along with three
other bottles in the back row (one labelled
ARQUEBUSADE is illustrated if fig. 3.6),
together with toiletry and writing materials,
all secured with a lock mechanism by Bramah
of 194 Piccadilly who held a Royal Warrant as
shown by the incised crown. The key survives.
The sides had sunken brass carrying handles.
Edwards was a manufacturer of Writing and
Dressing Cases by appointment to His Majesty
King George IV (fig. 3.7) and significantly to
the Duke of York amongst others. He was also
a repairer of Dressing cases and exchanged
damaged cases for new ones. He was a supplier
of glass, silver and fine cutlery. He used Allen
Dominy of 55 Red Lion Street, Holborn, as a
silversmith (Grimwade 27). So he made the
silverware for the three bottles and two flacons
in the back row, the large travelling black
ink-well (fig.3.8, and sander (or pounce pot)
in the next row, the small travelling copying
ink-well (fig. 3.8) and wafers container in the
next row and the toilet boxes in the front. This

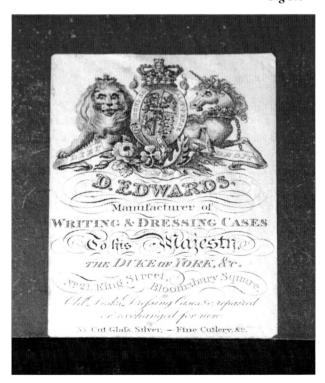

Fig 3.7

was an up-market case. In the lid a
self-standing leather backed tabbed
mirror lifts out. In front of this there is a pull-

Fig 3.8

Fig 3.9

Fig 3.10

down writing desk with letters and paper folder underneath. All this is concealed behind a decorative lid. The inks stand on a tray with lifting handles, which when removed reveals another tightly fitting tray which then when lifted out reveals a secret bottom tray which also lifts out (fig. 3.9). All the silver bottle tops are engraved with the owner's initials "CCA".

3.11 Another similar rosewood box had the two flacons arranged down the left-hand side. It

appears that they were mounted in silver by John Reily in 1816 but any labels that were accompanying are missing (3.12). In 1832 a brass bound rosewood campaign case contained five glass toilet water bottles mounted in silver gilt by Charles Reily and George Storer . The lift out tray had a silver scent bottle with engraving that matched other items in the box (3.13).

3.12 In France in 1840 Aucoc Ainé of 4 Rue de la Paix made a splendid necessaire (3.14) designed perhaps for a newly married couple with seven cut glass perfume bottles of various shapes and sizes (fig. 3.10). Suspension labels could easily have been employed.

3.13 An early Victorian leather-bound travelling case contained 22 bottles each approximately 9 cm. long with cork stoppers, some with hand-written and some with printed labels from a variety of chemists, which had a professional appearance (fig. 4.27). However, an apothecary's chest is quite different. It would have had at least 16 glass stoppered bottles along with a pestle and mortar for mixing, a mixing bowl and weighing scales, a separate compartment for poisons, all contained in a stout mahogany medical cabinet. An example is illustrated (fig. 4.26). Suspension labels would not have been employed in this kind of professional medical travelling box.

Fig 3.11

Fig 3.12

3.14 William Neal made in London in 1850 silver covers for six glass jars and four identical cut glass bottles in a fitted case with brass handles, containing a mirror in the lid and a writing facility in the base (fig. 3.11). Labels would have been required to distinguish the contents of identical bottles and jars (3.15).

3.15 The Royal Warrant holders Wells and Lambe of Cockspur Street in London in 1857 made a similar ebonised wooden box with three identical toilet water bottles and four circular glass jars all mounted in silver by James Vickery (3.16).

3.16 Halstaff and Hannaford of 226 Regent Street in London in 1860 made a similar brass bound wooden box (fig. 3.12) in burr walnut fitted with six identical cut glass bottles mounted by John Howes (3.17), two small pots and three boxes, with engraving, and with a concealed stationery wallet and mirror in the cover. It had two lift out trays with manicure items. A spring released a secret drawer for jewellery. Made in the same year for Aspreys of 166 Bond Street and 22 Albemarle Street, another ladies box made in coromandel contained a silver-gilt toilet set by James Vickery set off by blue velvet lining which included 5 screw top glass

jars, and a stationary folder and a detachable mirror in the lid. This was made for Elizabeth, Countess of Iddesleigh (3.19).

3.17 Thomas Wimbush of London made the mounts for 9 silver topped glass containers housed in a rosewood toilet case of 1852.

3.18 William Tween and Frederick Purnell in London in 1865 made mounts in silver-gilt for five square cut glass bottles in two sizes and for six cylindrical cut glass bottles in two sizes in a fitted rosewood case (fig.3.13) with brass fitments (3.18).

3.19 William Lund of Cornhill and 24 Fleet Street in the City of London was a noted box maker of his time. In 1871 he made a brass edged coromandel dressing case (3.19) lined with satinwood with two silver mounted bottles and two silver mounted jars. Its contents included a pen-knife, button hook, nail file, ear wax remover, cork-screw, hair brushes, clothes brushes and a large mirror.

3.20 Jenner and Knewstub of 33 St. James's Street in London in 1889 made a similar box in coromandel with six cut glass bottles in three sizes with silver mounts by the same makers

Fig 3.13

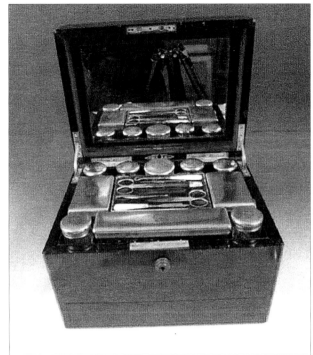

Fig 3.14

(3.20), and in 1899 Asprey's made (3.21) a less pretentious box again in coromandel with four cut glass bottles in two sizes with silver mounts together with two further cut glass bottles this time with silver-gilt mounts to match the three larger rectangular boxes which had silver-gilt lids (fig. 3.14).

3.21 With the onset of the Art Nouveau and Art Deco periods fashions changed and travelling cases became very stylish and were often called "Vanity Cases". The need for labelling clearly diminished. So, for example, a magnificent leather travelling case by Finnegans of London and Liverpool made in 1929, with a special canvas cover to preserve the leather, contained no objects capable of carrying a bottle ticket. A Mappin and Webb 18ct. gold mounted blond tortoiseshell toilet set of 1923 (fig. 7.11) contained a pair of shouldered square cut-glass toilet water bottles with stoppers, but their design with vacant cartouches precluded the use of suspension labels (3.22).

NOTES

3.1   See below, Chapter Five, The Perfume Labels.
3.2   Botanick Table, Bodleian Library MS Ashmole 1820b, folio 565.
3.3   "A travelling cordial case", 1 WLCJ 5, p 61.
3.4   Bonham's 25.11.2004, Lot 52.
3.5   Christies' 19.10. 2004, Lot 212.
3.6   Lecurieux, 22.10.2004
3.7   Christies' 19.10.2004, Lot 168.
3.8   Christies' 19.10.2004, Lot 155.
3.9   Christies' 19.10.2004, Lot 156.
3.10 Christies' 19.10.2004, Lot 162.
3.11 Christies' 19.10.2004, Lot 154.
3.12 Advertised for sale by an antique dealer in Chester, Connecticut,USA on

www.trocadero.com
3.13 Woolley's 21.4.2004, Lot 1744.
3.14 Lecurieux, reference NEA 0046.
3.15 Christies' 19.10.2004, Lot 211.
3.16 Woolley's 21.4.2004, Lot 1633.
3.17 Woolley's 28.7.2004, Lot 1178.
3.18 Christies' 19.10.2004, Lot 210.
3.19 Woolley's 28.1.2009, Lot 995.
3.20 Sworder's 6.2.2005, Lot 158.
3.21 Sworder's 26.11.2008, Lot 152.
3.22 Other examples of design precluding the use of labels are cited below in Chapter Five, The Perfume Labels.

INTERIOR OF AN APOTHECARY'S SHOP.
Late XIV. or Early XV. Century.   Flemish.
(From an Old Painting.)

Fig 4.1

Fig 4.11 Porcelain Apothecary's Jar circa 1850

CHAPTER 4

MEDICINAL LABELS

4.1 The earliest known labels with suspension chains are travelling box silver labels by Margaret Binley (1764 -1778) for VERBENA, a remedy for migraine, and AROMATIC VINEGAR, a remedy for faintness.

4.2 Apothecaries, the doctors of the day, followed by chemists and pharmacists, have for centuries labelled their jars and bottles. But the titles were built in or stuck on. Antiques Roadshow (4.1) has displayed a re-inforced octagonal metal campaign medical chest circa 1590-1610, with a carrying handle, containing 10 glass stoppered bottles with hand-written paper titles. One referred to "Senapod" which was a popular laxative. Another chest, a little later in date, has also been displayed by Antiques Roadshow (4.2) containing 14 bottles with their descriptions inscribed in ink, including a label for "Senapods". This chest was lockable. The Wellcome Library London, has an engraving of c.1630 showing the library (fig. 4.10) and laboratory of an apothecary. The drug jars and bottles on the shelves are all labelled but not with suspended labels which might drop off or be mixed up causing confusion. The Library also has an etching, with water colouring, by Henry Heath in 1825, showing the interior of a stylish pharmacy with the pharmacist serving a customer and an apprentice at work. Fig 4.1 shows an Apothecary's shop of circa

1400. The Society of Pharmacists of Bruges in Belgium have premises alongside and forming part of the complex of St. John's Hospital, one of the oldest in Europe, dating back to 1181 according to hospital rules and regulations of that date to be obeyed by the Brothers of St. John and the Augustinian nuns who tended the sick there. In 1778 Jan Beerblock painted a view of the main ward of the principal hospital building (fig. 4.2). The ward adjoins the richly decorated Shrine of St. Ursula. An interior walled herb garden, surrounded by cloisters, with a centrally placed well, leads off the passage to the pharmacy.

4.3 This Renaissance style pharmacy (fig. 4.3) contains some 26 shelves of labelled drug jars. The eleven blue and white glazed jars on the right hand side have intricately designed escutcheon shaped labels in blue with the following titles:

A. GRAMANIS	CORT.AURA
CANTAGIN	CORT.CITRI
CARDUTBENI	MATVIE
CICHOTEI	PENNICK WATER
CORD. AURAN	SAMBUC
COCHLEARD	

Two of the rather large jars stored at a lower level on the right hand side are inscribed respectively TINCT.CHIN.COMP. and

Fig 4.2

Fig 4.3

Fig 4.4

Fig 4.7

Fig 4.5

Fig 4.6

AETHER. SULP.ALCOOL. Some 20 box drawers are placed on top of the shelving on both sides of the room. The inscription on one box, by way of example, reads CHIRON. OPOPAN, G. RES. OPOPANAX (see further paragraph 9.399 below). The next room to the pharmacy is the Society's Court Room with oil paintings of previous Masters placed in serried ranks around the walls. A door leads into the next room, the Hall or Meeting Place for the membership. A further door then leads into the ambulatory or cloisters which lead back to the pharmacy. The general arrangements remind one of Apothecaries' Hall in London belonging to the Livery Company of that name.

4.4 The Museum of St. Bartholomew's Hospital has three early drug jars which reflect the standard of treatment. A Seventeenth Century imported "waisted style" jar (it allowed easy removal from a crowded shelf) for "Pill. di agarico" had its contents Pilulae di Agaricus added to it later. These were pills made from a fungus containing agaric acid used for treating night sweats or consumption (pulmonary tuberculosis). A jar with this title

was actually found on the St. Bartholomew's Hospital site in the City of London. A later lidded jar marked "PIL.RHEI.CO" would have contained Pilulae Rhei Composita being pills made from a substance obtained from rhubarb used as a purge. A third jar discovered there marked "UNG (POISON) BELLAD" would have contained the ointment Unguentum Belladonnae made from the leaves and root of the plant deadly nightshade containing atropine which would have been used as a muscle relaxant.

4.5 Two early apothecary's jars with pourers are illustrated with simulated hanging labels for the intriguing titles of E:BAC:SAM: and V:RUB: (fig. 4.4). London Delftware jars from about 1690 were popular with apothecaries. Illustrated in fig. 4.5 is "LOH:PVL:UVLP", also marked "E.R 1697", for Lohoch de Pulmone Vulpis which would have contained dried fox lung and in fig.4.6 "E:DIASCORD" which would have contained an Electuary of Water Germander known as Electuarium Diascorium made up of opium, herbs and spices in honey, sugar and Canary wine. It has been described as "A great strengthener of the stomach and bowels….a very mischievious way some nurses have got of giving their children this medicine to make them sleep". Eighteenth Century London Delftware jars have been shown in Antique Collecting (4.3), some 370mm high with covers for C:CYNOSB: and C:E:CORTEAU. Others are illustrated in fig. 4.7 for P:EX:2BUS: and for EX:CORT:PERUV:M: in simulated long narrow rectangular labels (4.4). The Wellcome Library, London, has early Italian faience handled jars for "AQUA.ROSATA"

Fig 4.8

and "AQUA.DE.BERTONICI" (fig.4.8). The Science Museum in London has pairs of handled ceramic jars for "THURIACA" and "SUGLOSSI" (fig. 4.9) and for "BATA UNGUENT" and "LEM:D'APPIO". Sworders sale on 20th September 2011 at Stansted Mountfitchet included an English Delftware wet drug jar labelled S:SUCC:LIMON for a syrup of lemon juice (4.5). Another pair of handled ceramic jars has been noted for U:RASINI and U:SOLIS, the letter U standing for UNGUENTUM being an ointment or salve. An unguentary dealt in perfumes and perfumed ointments. Nineteenth Century jars carry hand painted titles such as the porcelain jar of circa 1850 (fig. 4.11) for "N:d:PO E:POL:". The Hoffbrand jars collection at one time displayed in Apothecaries' Hall in the City of London has some interesting examples with a wide variety of titles, including "E.DIACASS.C.MANNA" for Electuarium Diacassiae cum Manna or PRUNES and "S: ZINZIB" for Syrupus Zingiber or GINGER. The Thackray Museum in Sheffield has an even larger collection of jars. Pilkington jars of the 1906-1914 period carry intriguing titles such as "PROP" and "POTIOR DAR BELLO", the work of William Salter Mycock. Some jars have pourers to aid the dispensing of V:RUB and E:BAC:SAM: (fig.4.4). A lidded Staffordshire jar beautifully painted "LEECHES" has been noted, the lid having breathing holes for the blood-sucking occupants of the jar. Leeches were applied using a tube (a superior silver example of 1830 with a glass liner inside has been noted). One gave blood without seeing the action.

Fig 4.9

Fig 4.10 This shows an Apothecary's Workroom, circa 1630. The labelling of the various jars perhaps inspired the design of suspended labels.

Fig 4.12

Fig 4.13

Fig 4.14

4.6   Bates and Hunt (or in television terms "Barber and Goodman"), chemists and druggists serving the community (fig. 4.12) from 9 High Street, Blists Hill, Ironbridge, (fig.4.13) is a reproduction of a Victorian druggists, grocers and chemists shop (fig. 4.14) selling patent and proprietary medicines. The four glass coloured carboys portray the four elements of the alchemist: green for earth; yellow for air; blue for water; and red for fire. The shop had many paper labelled bottles and jars containing apothecary's mixtures such as "PV. AC.BORIC" and "COCAIN. HYDROCH" which was additionally marked "POISON" in red. Brightly coloured corked glass bottles contained useful mixers such as "BITTERS" (fig.9.32) and "MINT" (fig.9.178) related to the preparation of certain remedies.

4.7   Ann Seedhouse, chemist and optician (fig. 4.15),

served the community from her pharmacist's shop (fig.4.16) in Brownhills, Staffordshire, from 1939 to 1979 following traditional lines, selling old-fashioned herbal remedies (after all the word "drug" is derived from the Dutch for "dried plant") from earthenware pots on her shelves. She exhibited two early eighteenth century mixing jars with rococo style blue scroll design labels for "V:RUB:" and "E:BAC:SAM:" thought to be made circa 1730 in Lambeth (fig. 4.4). They would look good in the Royal Pharmaceutical Society's museum in Lambeth.

4.8   Paper titles were affixed in the nineteenth century to glass bottles looking like broad rectangular shaped wine labels often with gold coloured borders. Examples are PLUMBI OXID of curved long narrow rectangular shape with a gold border on a Bristol blue glass bottle accompanied by a POISON label, R:R:TICIA with a gold border on a Bristol blue glass bottle, GOMA TRAGACANTO ENT. on an escutcheon style label with a gold border on a green glass bottle and HYORARO TANNAB on a rectangular shaped label on a green glass bottle. Illustrated are Tinctures for Digitalis (fig 4.17), Chlor-morphine (fig. 4.18) and Rhei (fig. 4.19) and Tinctures for CANNAB:IND (POISON) (fig. 4.20) and CAMHOR:CO (fig. 4.22) along with VIN: PORTENS (fig. 4.21). It was however in the home that suspended labels were used, not in the shop. Home medicines became available with the spread of knowledge. Many titles in the list below were not

Fig 4.16

Fig 4.15

Fig 4.17   *Fig 4.18*   *Fig 4.19*   *Fig 4.20*   *Fig 4.21*   *Fig 4.22*

exclusively used in the boudoir for medicinal purposes. Hence wines and spirits were often used for medicinal purposes.

4.9 In 1694 the author of "The Ladies Dictionary: Being a General Entertainment for the Fair Sex", identifiable (perhaps wisely) only by the initials "H.N.", pontificated about diet, fashion, courtship and marriage. The recipe for keeping slim was somewhat daunting: "take an ounce and a half of oyl of foxes, and of oyl of lilies, capons grease, and goose grease, each two ounces; pine, rosin, Greek pitch and turpentine, of each two ounces." One had to boil this mixture adding an ounce of oyl of elder. Once cool it could be applied as a kind of poultice "to the place that languishes, or does not equally thrive". No labels have been noted for oil of foxes, oil of lilies or oil of elder, but perhaps the use of these particular oils had ceased by the 1760s when the first silver medicinal labels were introduced.

4.10 Mrs Smith, in her eighth edition of the "Compleat Housewife" of 1737 costing five shillings, set out a wide range of some 300 recipes for medicines "viz: Drinks, Syrups, Salves, and various other things" to deal with "Distempers, Pains, Aches, Wounds, Sores etc." (fig. 6.2). She also listed some 26 home-made wines, nearly all of which were thought to have medicinally beneficial effects.

4.11 Extensive use of wines and spirits is made by Dr. William Buchan as recorded in his "Home Doctor" first published in 1769 which ran into no less than nineteen editions. Details of his recipes for medicines using wine and the rationale behind such use are available. Some of his recipes have been recorded in the Journal in an article by Dr. Trevor Woodward on the use of wine for medicinal purposes (4.6). A publication of the Wine Advisory Board of San Francisco states that recent research has "provided strong support for the medical uses of wine, as well as explanations of its pharmacological effects". Wine can be taken to relieve anxiety. Some of the wines listed by Dr. Buchan are as follows:

Anthelmintic Wine
Antimonial Wine
Bitter Wine
Ipecacuanha Wine
Chalybeate or Steel Wine

4.12 The Domestic House Book, published in 1826, edited by Dr. William Scott, contains upwards of a thousand selected prescriptions for the use of families. The frontispiece shows Dr. Scott with his family engaged in mixing medicines (fig. 4.23). Cruikshank delineated the family of this general practitioner as they were in 1820. A somewhat doubtful remedy for rheumatism is illustrated (fig. 4.24) which is indicative of the standard achieved at this time. The six liqueurs mentioned in the 1638 Charter granted to the Distillers' Company, much under the influence of its medical founder, are as follows:

CHARTREUSE   KUMMEL
BENEDICTINE   MARASCHINO
GRAND MARNIER   CURACOA

4.13 The 34 medicinal liqueurs listed by Dr. Scott in 1826 are, in his order, as follows:

EAU DIVINE
RATAFIA D'ANGELIQUE
RATAFIA D'ANIS
HUILE D'ANIS
ANISETTE DE BOURDEAUX
RATAFIA DE CAFFE
RATAFIA D'ECORCE D'ORANGE
RATAFIA A LA PROVENCALE
RATAFIA DE VIOLETTES
RATAFIA DE NOYAUX
RATAFIA DE BRON DE NOIX
RATAFIA DE GENIEVRE
RATAFIA DE FRAISES
ESCUBAC
RATAFIA DE QUOINGS
RATAFIA DE CACAO
RATAFIA DE GRENOBLE
RATAFIA DE CERISES
RATAFIA DE CASSIS
RATAFIA DE FLEURS D'ORANGE
HUILE DE VANILLE
VESPETRO
RATAFIA A LA VIOLETTE
FENOUILETTE DE L'ILE DE RHE
ELEPHANT'S MILK
RATAFIA DE BAUME DE TOLU
CITRONELLE
CRÈME DE BARBADES
CEDRAT
PARFAIT AMOUR
HUILE DE VENUS
MARASQUIN DE GROSEILLES
CRÈME D'ORANGE
CRÈME DE NOYAU

4.14 He advises that all the liqueurs are good stimulants but that some must be taken with care! He singles out NOYAU for special mention: "The late Duke Charles of Lorraine nearly lost his life from swallowing some "Eau de Noyau" (water distilled from peach kernals) too strongly impregnated. It contains Prussic Acid, on which its deleterious principle depends". By way of comparison according to the Encyclopedia published in 1902 the most highly esteemed liqueurs include:

ABSINTHE	CURACOA
ALLASCH	GOLD-WATER
ANISETTE	KIRSCHENWASSER
BENEDICTINE	KUMMEL
BITTERS	MANDARINE
CHARTREUSE	MARASCHINO
CHERRY BRANDY	NOYEAU
CRÈME DE CACAO	PARFAIT AMOUR
CRÈME DE CAFÉ	PEPPERMINT
CRÈME DE MENTHE	POMERANS
CRÈME DE ROSE	RATAFIA
CRÈME DE THE	TRAPPISTINE
CREME DE VANILLE	VERMOUTH

4.15 Leonard Raven-Hill's sketch for "Punch" in 1906 (fig.4.25) portrays a doctor at work in Edwardian times. The doctor is asking an elderly patient if he had taken a box of pills prescribed for him : the patient replied that he had found the box difficult to swallow! Family or domestic medical boxes were maintained and taken upon travels. A medical box in the Titanic Drawing Offices in Belfast (4.7) contains LAUDENUM, BORACIC

Fig 4.23

Fig 4.24

For a Rheumatifm, or Pain in the Bones.

TAke a quart of milk, boil it, and turn it with three pints of fmall-beer, then ftrain the pof-fet on feven or nine globules of ftone-horfe dung tied up in a cloth, and boil it a quarter of an hour in the poffet-drink; when it is taken off the fire, prefs the cloth hard, and drink half a pint of this morning and night hot in bed ; if you pleafe you may add white-wine to it. This medicine is not good if troubled with the ftone.

To

ACID, EPSON SALTS, SAL VOLATILE, BICARBONATE OF SODA, TURKEY RHUBARB, ESSENCE OF JAMAICA GINGER, TURPENTINE, HIPPO WINE, HAZELINE, CHOLATE, MAGNESIUM CITRATE, PURIFIED, SANATOL, MORPHINE and AMYLE NITRATE, being paper labels mainly affixed to stoppered glass bottles. The bottom drawer of the box contained a needle, syringe and notebook. LAUDENUM was in red letters being a poison. Some professional medical boxes had a secret compartment in the back for dangerous drugs. In this case the AMYLE NITRATE was contained in a separate box within glass capsules. The medical box was supplied circa 1860 by F. Jones MPS of Liverpool. Another example is illustrated in fig. 4.26. Silver labels suspended by a chain would not have been employed in this kind of professional medical travelling box, which would have contained at least 16 glass stoppered bottles. On the other hand boudoir labels exist for medical remedies such as LAUDANUM, POISON LAUDANUM, BORACIC ACID, SAL VOLATILE, BICARBONATE OF SODA, ESSENCE OF GINGER and MAGNESIUM CITRATE. So household medical chests and travelling medical chests were perhaps kitted out with bottles bearing silver labels. Some boxes contain a pestle and mortar or a mixing bowl and weighing scales. Some are light weight for easy travelling such as the 22 bottle set contained in a leather travelling case (fig.4.27). Each bottle is some 9cm long and has a cork stopper. Some have hand-written and some have printed labels. The printed labels come from a variety of chemists. Thus suspended labels would not have been found in Dr. R.J. Church's smart tooled leather attaché case of around 1900 in date which contained five cylindrical tube holders in the underside of the lid and compartments for four large bottles, three small bottles and three circular small pots, together with a lidded rectangular storage area for various instruments. One of the large bottles was missing. The remaining three had adhesive paper labels for URA WARBUS (two teaspoonfuls), SPIRIT OF NITRATE and SAL VOLATILE. The three smaller bottles were labelled ORE OF ALOE (10-30 minutes), OIL OF IPECAC (2 grains to 30 grains) and some handwritten concoction. All these remedies were supplied by Dr. Arrowsmith, a Pharmaceutical Chemist, from a dispensary at 3 High Street Broadstairs in Kent.

4.16 Medicinal labels adorned bottles or decanters in various locations. The bedroom

Fig 4.25

Fig 4.26

probably had liquids to help one sleep such as CHLOROFORM, the relaxant CLARY, COWSLIP, GIGGLEWATER and MUM, alongside the refreshing WATER, SURFEIT WATER, TAR WATER and GOUTTES DE MALTHE. The dressing room had a range of bottles on the dressing table, mainly for perfumes. The beautiful small labels for ACID in both silver and mother of pearl may have been prominent, in easy range to deal with flatulence. EAU ASTRINGENTE could have helped with the complexion.

Fig 4.27

4.17 Bottles held quantities of smelling salts from which to fill up the flacons in the handbag or the vinaigrettes or the small bottles in the travelling case. The tiny labels for VERBENA and AROMATIC VINEGAR by Margaret Binley which could be as early as 1764 are probably the earliest known silver boudoir labels. So Benedict, CAMPHORATED SPIRITS, GINGER, EAU D'ALIBURGH, Hartshorn, and SAL VOLATILE could all have been used in later travelling boxes. Larger enamel, porcelain and silver labels could have adorned the larger storage bottles, such as enamels for AMHONIA and AMMONIA. In the gentleman's dressing room there could be housed his campaign travelling box with EAU DE MIEL, ARQUEBUSADE and BALM for use during the time of the Napoleonic wars and subsequently (1790 -1830). A cupboard in the nursery could have contained BARLEY WATER, COLLICK WATER and perhaps MILTON and SOURING. It would be handy to have ANGELICA, BENZOIN, BICARBONATE OF SODA, SODA, BORACIC, BORIC, CASTOR OIL, ELIXIR, CINNAMON,CLOVES, CITRATE MAGNESIUM, EAU BORIQUEE, GARGLE, GLYCO-THERMALINE, GRAIN, GRAIN DE LIN, HUILE DE VENUS, HUNGARY WATER, IODINE, LEMON, LEMON BALM and METHYLATED SPIRITS in the bathroom, together with POTASH which was used in the bath.

4.18 In the rest of the boudoir one might find a side table with delightful cordials used as aids for digestion such as AMEIXORIA, ANIS, ANISEED, AQUA MIRABEL, ELDER,

FONTINIAC, GARUS, KIRCHEN WASSER, LEDSAM VALE, GENTIAN and ORGEAT. One would also have handy a range of pick-me-ups such as BITTERS, BRANDY, COOKING PORT, CAMPHOR JULEP, COGNAC, CORDIAL GIN, FINE NAPOLEON, GOMME, GUM, GUM SYRUP, CRÈME DE NOYEAU, CRÈME DE THE, GINGER, LEMON CHAREC, LEMON BRANDY, LOVAGE, ORANGE, NOYEAU, RASPAIL, RASPBERRY VINEGAR. RED VINEGAR and VINEGAR.

4.19 One would still perhaps need to go to a pharmacist for a cure-all, often claimed to be based upon a secret remedy which may have included ground rhubarb root, liquorice, soap powder as a laxative, and syrup. The British Medical Association's publication on "Secret Remedies" shows that people in mid-Victorian times were most concerned about the spread of diseases, through over-crowding, such as cholera, tuberculosis, influenza, typhus, typhoid and measles. The book contains detailed remedies, even "Pink Pills for Pale People". One could purchase a pills machine made by the pharmacist containing herbal remedies. Pills were actively marketed by Beecham, Holloway and Morison. They could be kept in labelled storage jars. The chemist could supply Dr. Thomas' Electic Oil for coughs, perhaps labelled "OIL", or Warner's safe cure for kidney and liver complaints, perhaps labelled "ELIXIR" or Dr. Jacob Townsend's "Sarsaparilla" for purifying the blood.

4.20 Spirits were used as sleep inducing pain killers or night caps. Small drinking vessels used for such purposes are known from the sixteenth century. One example dated 1590 contains one-eighth of a fluid ounce. A century later such small drinking vessels contained a quarter of a pint. An example of 1693 is called a noggin. A century later spirits were taken with hot water and sugar. A vessel dated 1786 to contain this mixture was called a toddy. So the habit of taking a night cap of BRANDY was seised upon by George Unite of Birmingham to produce in 1844 a vessel fit for the purpose. He made a small silver jug with handle and spout and stopper, capable of taking a dram, which became known as a BRANDY tot. After the Boer Wars and World War I injured officers were often given a tot or dram of WHISKEY or more rarely BRANDY or KUMMEL contained in individual small glass drinking vessels called toddies or noggins to which hot water and sugar were added. These vessels had individual suspended silver labels with the title in script with a flourish (fig. 4.28). This then became a popular night-cap to have in the boudoir. Examples have been noted by Samuel Jacob (1911), Atkin Brothers in Sheffield (1912), Hukin and Heath in Birmingham (1921, 1929 and 1936), and Heath and Middleton (1924). The WHISKEY tots could be topped up from a storage container often shaped in the 1900s as a whiskey barrel, or from a whiskey decanter, such as a Kingsware Royal Doulton Dewar's Whiskey jug of 1911, for example. Small quantities of BRANDY could also be drunk from a glass known as a balloon. These small amounts were known as snifters, and often said to be taken for medicinal purposes.

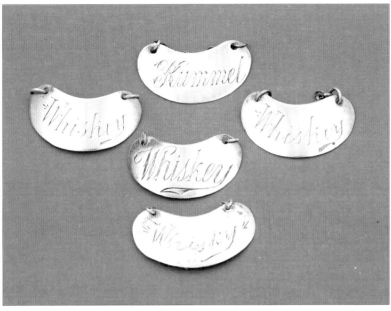

Fig 4.28

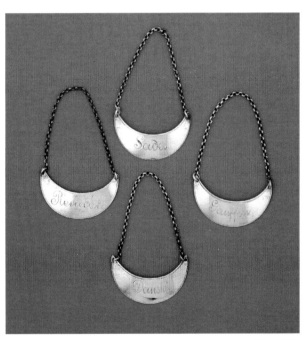

Fig 4.29

4.21 In Sweden one could have a choice of Saides, Gouffin, Renadt or Danski (fig.4.29) as rectified or purified spirits such as those made by Lars Olssen Smith from about 1877 onwards as varieties of vodkas, aquavits or snaps. These were not necessarily medically approved. Indeed some cultures in harsh climates ban their use altogether. In Denmark and Sweden

however it is customary for snaps, being small shots of strong alcoholic beverages, to be drunk at mealtimes accompanied by some ceremony. In the USA schnapps (from the German schnaps) ranks as a liqueur because of the added sugar content.

4.22 A full list of titles of medicinal boudoir labels is set out below. Many titles in the list were not exclusively used in the boudoir for medicinal purposes. Many could have been used elsewhere to label a wine or spirit decanter.

MEDICINAL LABELS

A BRANDY
ACID
AMEIXORIA
AMHONIA
AMMONIA
AMONIA
ANGELICA
ANIS
ANISEED
ANISETTE
Anisette blanc
ANNESEED
ANNICETTE
ANNISCETTE
ANNISEED
ANNISETTE
AQUA MIRABEL
ARISETTE
AROMATIC VINEGAR
ARQUEBUSADE
ARQUEBUZADE
AURQUEBUSADE
BALM
BARLEY WATER
Benedict
BENZIN
BENZOIN
BENZOIN LOTION
BENZOIN SALTS
BENZOIN SOLUTION
BICARBONATE OF SODA
BICARB SODA
BICARD DE SOUDE
BITTER
BITTERS
BORACIC
BORACIC ACID
BORACIC LOTION
BORACIC WATER
BORIC
BORIC ACID
BRANDY
C.BRANDY
C,BRANDY
C. PORT
CAMPH DROPS
CAMPHtd SPIRIT
CAMPHOR JUICE
CAMPHOR JULEP
CAMPHORATED SPIRITS OF
    WINE
CAPILARE
CAPILLAIRE
CAPISCUM
CASTOR OIL
CHARTREUSE

CHERRY
Chloroform
CINNAMON
CINQ A SEPT
CITRATE MAGNESIA
CITROEN
CITRON
CLARY
CLOVES
COGNAC
Cogniac
COLLICK WATER
Cooking Brandy
Cooking Sherry
CORDIAL
CORDIAL GIN
CORDIALS
COWSLIP
COWSLIP WINE
CRÈME DE NOYAU
CRÈME DE NOYEAU
Crème De Noyeau
Crème De Thé
DAMSON
Damson Cordial
DANDYLION WINE
D'ORGEAT
EAU ASTRINGENTE
EAU BORIGUEE
EAU BORIGUEES
EAU BORIGUES
EAU BORIQUEE
EAU D'ALIBURGH
EAU D'ARQUEBUSE
EAU de MIEL
EAU DE NOYAU
EAU DE NOYAUX
EAU DE NOYEAU
EAU D'OR
EAU. D'OR
Eau de Vie
ELDER
ELDER.FLOWER
ELDER FLOWER
Elderflower
Elder flower
ELDER VINEGAR
ELIXIR
ELIXIR DE SPA
ESSENCE GINGER
ESSENCE OF GINGER
ESS GINGER
EYE
EYE DROPS
EYE LOTION
EYE WASH

FINE NAPOLEON
FINENAPOLEON
FONTANIAC
FONTINIAC
French Vinegar
FRENCH VINEGAR
FRONTINIAC
FRONTIGNAC
FRONTIGNAN
FRONTIGNIA
GARGLE
GARUS
G BRANDY
Gentiane
GIGGLEWATER
GINGER
GINGER BRANDY
Ginger Wine
GINGERETTE
GLYCO-THYMOLINE
GOLD WATER
Gomme
GOUTTES de MALTHE
GOUTTES DE MALTHES
GRAIN
Graine de lin
Grandad's Medicine Bottle
GULDEWATER
GUM
GUM SYRUP
HARTSHORN
Hartshorn
HUILE
HUILE D
HUILE. D'ANIS DES INDES
HUILE D'ANIS ROUGE
HUILE DE VENUS
HUILE DE VINUS
HUILLE DE ANIS
HUILLE DE VENIS
HUILLE DE VENUS
HUNGARY
HUNGARY WATER
IODINE
JAM. VINEGAR
Kersevaser
KHOOSH
KIRCCH WASSER
KIRCHEN WASSER
KIRCHEVASSER
KIROCHVASSER
KIRSH A WASSER
KIRSCHENWASER
KIRSCHENWASSER
KIRSCHWASSER

MEDICINAL LABELS

KIRSCH-WASSER
KIRSHWASSER
LACHRYMA
LACKRYMA
LACRIMA
LACRYMA
LAUDANUM
L'EAU CHAREC
LEDSAM VALE
LEMON
Lemon Balm
LEMON BRANDY
LIMOSIN
LOVAGE
MAGNESIUM CITRATE
MAPLE SYRUP
METHELATED SPIRIT
METHYLATED
METHYLATED SPIRIT
METHYLATED SPIRITS
MILTON
MINT
MUM
NOIEAU
NOYAU
NOYEAU
NOYEO
NOYEU
OIL
OPPOPONAX

ORANGE
ORANGE BITTERS
ORANGE CURACAO
ORANGE FLOWER
Orgeat
PAFAIT AMOUR
PARFAIL AMOUR
PARFAIT AMOUR
PARFAIT D'AMOUR
PARFAITE AMOUR
PARSNIP
PEACH
PEPPERMINT
PICK-ME-UP
POISON
POISON LAUDENUM
POTASH
PRUNES
P.MINT
P.VINEGAR
RASPAIL
RASPBERRY VINEGAR
RATAFIA
RATAFIE
RATIFEE
RATAFIER
RATIFIE
RATIFIA
RATTAFIER
RATTFIER

R. CONSTANTIA
RED CONSTANTIA
R. VINEGAR
RED VINEGAR
ROSEWATER
SAL VOLATILE
SCYRUP
SODA
SODA WATER
SOURING
SPIRIT LAVENDER
SPRUCE
SULTANA
SURFEIT WATER
S.WATER
TAR WATER
TISSANNE
TOAST
TONIC
VERBENA
VERVEINE
VERVIENE
VIELLE FINE
VIELLEFINE
VINAIGRE
Vinaigre Ordinaire
VINEGAR
W. CONSTANTIA
WHITE VINEGAR
YLANG-YLANG

NOTES

(4.1) Antiques Roadshow, 20.12.2010. On 24.1.2012 the programme Flog It! recorded a visit to Crathes Castle in Aberdeenshire where a pharmacy cabinet containing 120 drawers was displayed. Titles included:-
ALUMNI SULPHUR
CERA ALB
GUM SHELLAC
LIG BASSAFR
M. LEAVES
PULV. VIOLET
RAD DOBA
RAD SARSAE
RAD SCILL
SUPPOSITORIES
The cabinet later on was sold at auction for £3,600.

(4.2) Antiques Roadshow, 25.7.2010

(4.3) Antique Collecting, July/August 2004

(4.4) Antique Collecting, July/August 2010

(4.5)   Sworders sale, 20.9.2011

(4.6) 4 WLCJ 8, p151

(4.7) Antiques Roadshow, 12.12.2010. The 100th anniversary of the sinking of the Titanic was marked by the opening of the new museum called the "Titanic Belfast" in April 2012 in Northern Ireland.

Fig 5.1

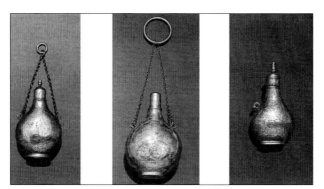

Fig 5.2

CHAPTER 5

PERFUME LABELS

5.1 The word "perfume" comes from the Latin, per meaning "through" and fumus meaning "smoke". So perhaps its earliest form was incense, the burning of sweet herbs and spices at various ceremonies. The Mesopotamians, Egyptians, Phoenicians, Etruscans, Greeks and Romans all had perfumery vessels. A collection of these can be seen in the Museu del Parfum in Barcelona. The Greeks used one-handled perfume vases. Queen Hatshepsut of Egypt was keen on aromatics and her subjects carried perfumes with them. Queen Cleopatra was lavish in her use of perfumes. The Israelites were instructed by Moses on how to make perfumes. King David infused his clothes with aloe and cassia. The making of the first liquid perfume is attributed to the Greeks, mixing oils with powders. So for example olive oil or almond oil were mixed with powdered lilies or roses. It was however Avicenna, an Arab chemist, who produced Rose Water by extracting oils from roses by means of distillation. The Roman Emperor Nero was said (in 54 AD) to have had a penchant for the scent of roses, which he used at banquets. Even to this day the Rose Water bowl is passed around at Livery Company dinners.

5.2 In the Sixteenth Century the travelling box contained perfume bottles which were not designed to take suspended labels. An example by Nicolas Schmidt with two perfume bottles and a perfume flask is illustrated in fig. 5.1. In the Seventeenth Century in Europe silver scent flasks were worn on a chain so as to be handy for use as an air and body freshener to mask other aromas that might be encountered. Circa 1608 a silver box decorated with floral enamels attached to a chain contained a variety of substances designed to mask unpleasant smells from rubbish thought to promote disease. Three examples of perfume flasks are illustrated (fig. 5.2) of compressed baluster form, dateable to 1680-1690. One has a finger ring attachment. In the early Twentyfirst Century ladies are still using fragrance pendants, hung in this case around the neck. Illustrated is a fragrance pendant utilising Green Adventurine, a semi-precious stone which can be filled with a few drops of perfume or essential oil through the use of a pipette (fig. 5.3). The Green Adventurine is believed to promote a feeling of well-being (5.1).

5.3 Delicate perfumes were produced in France, particularly at Grasse, and the French Court became "le cour parfumee". Even the hand-held fans were perfumed. Napoleon was a heavy user of cologne. Today perfumes are complex combinations of natural products such as essential oils and synthetic products to increase the intensity of the smell and make it last longer.

5.4 Perfumes come in various strengths. Alcohol is used as a liquid base for perfume. A basic but not exhaustive classification would be to identify three basic strengths. The purest, strongest and most expensive, having around 15 to 40% ratio of alcohol to perfume concentrates, is called Extrait, Extract or just Perfume . The less expensive but still long lasting, containing around 7 to 15% ratio of alcohol to perfume concentrates, is called Eau de Parfum. The cheapest and lighter, not lasting long, containing around 1 to 6% ratio of alcohol to perfume concentrates, is called Eau de Toilette, Eau Fraiche or Eau de Cologne.

Fig 5.3

5.5   Perfumes are added to spirits to create liqueurs. This causes confusion in identifying labels because sometimes the liqueur is given the name of the perfume. The Encyclopaedia published in 1902 defines liqueurs as "perfumed and sweetened spirits prepared for drinking, and for use as flavouring material in confectionery and cookery" (5.2). Balm leaves and tops are the principal ingredients of CHARTREUSE. Orange peel is the basis of CURACOA. Raspberry juice and orange flower water are used in the preparation of MARASHINO which is distilled from fermented cherry juice. Star anise, angelica root and orange peel are used to prepare KUMMEL, Allasch or Doppel-Kiimel. So liqueurs are often given confusing names such as NOYEAU, PEPPERMINT, CHERRY, MANDARINE, PARFAIT AMOUR, CRÈME DE VANILLE, CRÈME DE ROSE, CRÈME DE THE, CRÈME DE CAFÉ, VANILLE, RATAFIA, ANISCETTE and KIRSCHENWASSER. Perfumes are also added to flavour cocktails. An example would be POUSSE L'AMOUR.

5.6   Perfumes are added to toiletries such as bath oils, smoothing creams, body lotions, bath powders, soaps and body moisturisers. Perfumes were originally stored in alabaster, glass and porcelain containers. Not all perfume bottles were designed to have suspension

Fig 5.4

labels but those comprised in a perfume box or travelling case often had suspension labels so that they would not lose their identity when taken out and placed on a dressing table. Some perfume bottles are highly artistic and comprise a separate study area of collectibles. Brosse of Boulevard Boudon in Paris are suppliers of some of the finest perfume bottles in the world. Some gain immensely from having a silver-gilt suspended label on them. Labels look good hung on stand alone silver mounted scent bottles for use on the dressing table. As new fragrances were encountered new labels were the easiest means of recording the change using the same beautiful containers. To save expense slip labels became popular.

5.7   The Falcon Glasshouse of Apsley Pellatt during the years 1831 – 1837 produced some wonderful bottles in patented pressed and cut lead glass which so well displayed a bust of Queen Adelaide, but labels were still needed to identify the contents.

5.8   Paper labels competed from about 1860 with titles such as Alcool Pur de Lavande, Parfum aux fleurs, Essence de Fleurs, Eau de Cologne, Eau de Fleurs d'Oranges and Eau de Parfum. Paper labels were not used on silver mounted perfume bottles which varied in number from three to nine in Asprey's Dressing Boxes or Cases in use from the 1890s to the 1930s which looked stunningly beautiful when made of crocodile skin or coromondel. A glass and gilt perfume bottle, housed in an individualistic travelling case, made in 1873 by Samson Mordan, inscribed in 1875 CHIEN SHENG to commemorate the survival of a British built gunboat after the battle of Fuzhou at the beginning of the Sino French war, is an example of attached rather than suspended labelling (5.3).

5.9   Two scent bottles together with a funnel (fig.5.4) were a Christmas present to Queen Marie Antoinette of France (1755-1793) contained in a book-shaped leather case with the Queen's coat of arms on in gold (5.4). No labels are to be seen. Lockable tooled leather perfume bottle cases became popular.

Fig 5.5

5.10 Two scent bottles, with silver mounts made in Birmingham in 1936, which packed neatly into a snug leather case secured by a flap, comprising one bottle with a tapered interior and one bottle with a cylindrical interior but which were otherwise identical, were featured in a "Flog It!" TV programme (5.5). Since the glass bottles were a pair, the slightly different designs enabled one to be distinguished from the other without recourse to labelling. A similar art-deco pair of bottles taken out of their leather case is illustrated in fig. 5.5. The stoppers are colour coded, so labels were not required. A similar set of four bottles in a leather case has been noted. Each bottle was made of a different colour glass (blue, green, red and white) to aid identification. Perfumes were presented in wooden, ceramic and silver containers as illustrated in fig. 5.6. A mid-Victorian three bottle perfume set, each cut glass bottle being of triangular shape with a curved baseline, rather like a fan, could have been labelled. Each bottle had a mounted silver collar with worn hallmarks. The stand however was electroplated and unmarked.

5.11 A typical toilet box of the 1830s made by Savory and Moore (1821-1839) has been noted, containing in the back row a square-shaped glass bottle with a silver bottle label for LAVENDER attached in one corner and a square-shaped glass bottle with a silver bottle label for VIOLETTE attached in the other corner, with three circular cosmetic pots in between. In front were lift out trays for pins, safety pins, cottons, buttons, and such like. In the centre was a lift out tray for a nail file, buffer, button-hook and various other items.

5.12 Another typical toilet box made perhaps in 1857 by Wells and Lambe of Cockspur Street, who held Queen Victoria's Royal Warrant, was in ebonised wood and brass bound. It contained three toilet water bottles, three oblong glass boxes, four circular glass jars, a mounted ink-well, a mounted vesta holder, a mounted glass powder box with screw-down cover, a lift out steel manicure set with mother-of-pearl handles and other items. It had a lockable jewellery drawer. It also had a concealed stationery wallet and mirror. The silversmith was James Vickery (5.6)

5.13 In the mid Nineteenth Century a silver-gilt travelling perfume set made in France

Fig 5.6

Fig 5.7

Fig 5.8

Fig 5.9

comprised comprised a funnel of plain circular form and four scent bottles mounted in silver-gilt. The inside of the shagreen box had a mirror inset (5.7). Another box also made in France housed three scent bottles mounted in silver-gilt (5.8).

5.14 The Marshall Collection has EAU DE COLOGNE, ESSENCE OF ROSE and LAVENDER numbered 334-336 which Mrs Marshall noted were contained in a perfume box made circa 1815. A square shaped cut glass perfume bottle is illustrated bearing a silver label for the very popular EAU DE COLOGNE made by Samuel Hayne and Dudley Cater in 1841 (fig.5.7).

5.15 In the Twentieth Century an oriental silver-gilt box with inset gems was converted into a perfume box by Charles Dumenil so as to contain four silver-gilt mounted glass bottles with gem set covers in 1907 (5.9). The bottles could have been labelled. Perfume bottles in sets were housed in a variety of containers, including wooden, porcelain and silver as

shown in fig.5.6. Perfume bottles were taken out of their travelling box and displayed on the dressing table. They were often placed on trays made of silver or of glass as illustrated in figs. 5.8 and 5.9.

5.16 D.R. Harris & Co. Limited in the centre of club-land at Number 11, now number 29 , St. James's Street was established by Daniel Rotely Harris in 1790 as chemists and perfumers. The firm supplied essential oils such as GRAPEFRUIT, JASMINE, LAVENDER, LEMON, ORANGE, PEPPERMINT, ROSE and ROSEMARY. The firm was the inventor around 1850 of the original PICK-ME-UP. It is a supplier of mouthwashes and shaving brushes and creams. Speciality lines include ROSE WATER and BALM and various kinds of bath oils and essences. Eau de Portugal is still available as a hair lotion. Among the perfumes various toilet waters can still be supplied such as COLOGNE, LAVENDER WATER, ROSE BOUQUET, JOCKEY CLUB BOUQUET and ENGLISH

BOUQUET. The firm has held various Royal Appointments (5.10).

5.17 Another well-known perfumer is Geo. F. Trumper of 9 Curzon Street Mayfair and 1 Duke of York Street leading out of St. James's Square. This firm is particularly known for its ROSEWATER still in much demand. Taylor of Old Bond Street, a family run business, now of 74 Jermyn Street, specialise in shaving creams, soaps, gels and oil, using almond, lavender, lemon & lime and other flavourings.

5.18 Another well-known perfumer is J. Floris Limited of 89 Jermyn Street which was established in the Eighteenth Century as perfumers and manufacturers of toilet preparations. Of interest and relevance is the Company's small museum and the fact that the cabinets it used for the Great Exhibition of 1851 are still in use and on show.

5.19 During the Regency period chinoiserie was a great attraction, probably due to the fact that the Prince Regent adored all things French. The popularity of chinoiserie was reflected in the choice of design of French enamel perfume labels for MON PARFUM and EAU DE TOILETTE, which adorned perfume bottles standing on the Dressing Table top, sometimes standing on silver or glass trays (modelled perhaps on the earlier spoon trays) holding two or more bottles. William Aitken of Birmingham made a superb art nouveau tray in 1909 which housed scent bottles as well as brushes and combs.

5.20 Perfume bottles were often displayed in pairs on a glass tray such as those shown in fig. 5.8 and the more modern Aqua Allegorica and Lilia Bella shown in fig. 5.9. The late Victorian/Edwardian dressing table arrangement shown in fig. 1.10 illustrates five silver-mounted cut glass perfume bottles, all with stoppers, made by George Nathan and Ridley Hayes of Birmingham but hall-marked in Chester in 1906 (LAVENDER WATER), Judah Rosenthal and Samuel Jacob of London trading as Rosenthal, Jacob & Co in 1884 (ROSE WATER), Edwin Martin Thornton, a specialist glass cutter and mounter of London in 1883 (EAU DE TOILETTE), James Deakin & Sons of Sheffield in 1887 (EAU DE COLOGNE) and Horace Woodward Limited of Birmingham in 1900 (EAU DE ROSE), together with one unmounted square cut glass bottle bearing an enamel label for ELDER FLOWER WATER. Also shown are a jewel-box by Joseph Gloster of Birmingham in 1922 and a hand-mirror by Hayes & Co also of Birmingham in 1913. Also worthy of note is the fact that William Hutton and Sons in London mounted in silver in the 1890s rounded crystal cut-glass perfume bottles. The dressing table was as important for the boudoir label as the decanter was for the wine or spirit label and the soy frame for the sauce label. Gawsworth Hall, for example, contained an elegant dressing table complex, thus setting the scene for the display of boudoir labels.

5.21 In August 1957 the "Times" recorded one of the monks of the Cistercian community of Caldey Island off the Pembrokeshire coast measuring the ingredients of one of their famous perfumes (fig. 5.10) and other monks (fig. 5.11) picking lavender to make the island's LAVENDER toilet water (5.11).

5.22 The dressing table arrangement shown in fig. 14.1 (see also fig. 1.10) shows a carafe labelled with an enamel MOUTH WASH (toiletry) with copper chain, an enamel BORACIC (medicinal) on a medicine bottle,

Fig 5.10

Fig 5.11

a Crown Staffordshire porcelain FRENCH (perfume) on an elegant perfume bottle and a silver CORDIAL (soft drink) on a flask by Robert Gray of Glasgow by way of example of glass containers used.

Fig 5.12

5.23 Since 1623 glassmakers Verreries Pochet et du Courval, of 121 Quai Valmy in Paris, have been making glass bottles, stoppers and jars or flaconnages for prestige fragrances and cosmetics such as Guerlain's "Shalimar" (5.12). One of their prolific productions is illustrated (fig. 5.12) marked "DE ROSE" for Eau de Rose or Esprit de Rose and "HP" for Pochet. Bottles were designed for a variety of perfumes and toiletries, upon which enamel labels were displayed, such as EAU DE COLOGNE as in illustration fig. 5.12 upon a bottle with distinctive shape.

5.24 Set out below is a list of Perfume Titles found on suspended labels.

Acqua di Parma
ACQUA D'ORO
ALBA
ALBAFLOR
ALBA FLORA
ALBA-FLORA
ARBAFLOR
ATTAR OF ROSES
BENZOIN
BENZOIN LOTION
BENZOIN SALTS
BENZOIN SOLUTION
BERGAMOT
BOUQUET
BOUQUET DE ROI
BOUQUET DU ROI
C.De ROSE
C.DE ROSE
Carnation
CELLADON
CHIPRE
CHYPRE
CINNAMON
CITROEN
CITRON
CITRONELLE
CLOVES
COLARES
COLLARES
COLOGNE
CRÈME de CAFÉ

CRÈME DE NOYAU
CRÈME DE NOYEAU
Crème De Noyeau
CRÈME de NOYEAU ROUGE
CRÈME DE PORTUGAL
CRÈME DE. PORTUGALE
CRÈME DE PORTUGAU
CRÈME DE ROSE
CREMEDEROSE
CRÈME DE VANILLE
CRÈME DE VIOLETTES
DACTILLUS
DACTILUS
Delicieux
EAU·DE·CAFFÉ
EAU DE CELADON
EAU DE CELLADON
EAU DE COL.
EAU DE COLODON
EAU DE COLOGNE
EAU.DE.COLOGNE
Eau de Cologne
Eau de Cologne ROSSE
EAU DE LAVANDE
EAU DE LUBIN
EAU de MIEL
EAU.DE.PORTUGAL
EAU DE PORTUGAL
EAU DE ROSE
EAU DE TOILETTE
EAU DE TOILETTE VERVEINE

EAU DE VIOLETTE
EAU D'OR
EAU. D'OR
EAU d'ORANGE
EAU D'ORANGE
ELDER
ELDER FLOWER
ELDER.FLOWER
Elderflower
ELDER FLOWER WATE
ELDER FLOWER WATER
ELDER-FLOWER WATER
ELDer FLOwer WATer
ELDR. FLOR. WATER
ELDr FLOr WATER
ELIXIR
ELIXIR DE SPA
ESPRIT DE ROSE
ESS BOUGET
Ess Bouquet
ESS BOUQUET
ESSENCE
ESSENCE BOUQUET
ESSENCE OF ROSE
Eyes
FINE NAPOLEON
FINENAPOLEON
FLORIDA WATER
Florida Water
FONTENAY
FOUR BORO'

PERFUME LABELS

FL. D'ORANGE
FLEUR D'ORANGE
Fleurs d'Orange
Floral Lilac Toilet Water
Fr. d'ORANGE
FLORIS VERVEINE
FRANCIPANNI
FRANGIPANI
FRANGIPANNI
FRANNGIPANI
FRENCH
French
GRANDE MAISON
H
HAMMAM BOUQUET
HAY
HUILE
HUILE.D'ANIS DES INDIES
HUILE D'ANIS ROUGE
HUILE D'ORANGE
HUILE de ROSE
Hui.le de Rose
HUILE DE VANILLE
HUILE DE VENUS
HUILE DE VINUS
HUILLE DE ANIS
HUILLE DE ROYAUX
HUILLE DE VANILLE
HUILLE DE VENIS
HUILLE DE VENUS
Infusion d'Homme
JASMIN
JASMINE
JESSAMINE
Jockey Club
JONQUIL
LA
LAIT D'IRIS
LA ROSE
LAROSE
LAVANDE

LAVENDAR
LAVENDER
LAVENDER WATER
Lavender Water
LILAC
Lily of the Valley
Limoux
LOCH LOMOND
L.W.
Milk of Roses
MILLE FLEURS
MILLE FLUERS
MINT
MIXT
MON PARFUM
MOSS ROSE
MURON
MUSCAT
MUSCATE
NEW MOWN HAY
NOIEAU
NOYAU
NOYEAU
ODC
O.D.C
ODV
O.D.V.
ODV de Danzic
OIL
OPPOPONAX
ORANGE
ORANGE FLOWER
PAFAIT AMOUR
PARFAIL AMOUR
PARFAIT D'AMOUR
PARFAITE AMOUR
PERFUME
PORTUGAL
POUSSE L'AMOUR
Pt. George
RATAFIA DE FLEUR D'ORANGE

RATAFIAT
Ratafiat de fleur d'orange
RONDELETIA
ROSA
ROSE
Rose
Rosebud Perfume
ROSE GERANIUM
ROSEMARY
ROSE WATER
ROSE.WATER
ROSE-WATER
ROSEWATER
ROSE WATER & GLYCERINE
SAL VOLATILE
SAVRON
S. WATER
SCUBAC
SEVE
SPIRIT LAVENDER
Star Mise
SWEET
Sweet Pea
VANILLE
VERBENA
VERVEINE
VERVIENE
VIELLE FINE
VIELLEFINE
VIOLET
VIOLET DE PARME
VIOLETE
VIOLETTE
VIOLETTE DE PARME
VIOLLETE
WHITE LILAC
White Lilac
WHITE ROSE
WOODBINE
YLANG YLANG

NOTES

(5.1) Following the success of the pomander with ladies, gentlemen adopted the vinegar stick (sometimes called the physician's cane being often carried by doctors) which was thought to have medicinal benefit, and then the vinaigrette, the sponge in which was often soaked in aromatic vinegar: see further Alexis Butcher, "Making Scents of the Vinaigrette", Antique Collecting, February 2012.

(5.2) Http://www.1902encylopaedia.com.

(5.3) Illustrated in J.H. Bourdon-Smith's catalogue number 49, Autumn 2011, p36.

(5.4) Reference number 0430A in the Museum collection administered by the Planas Giralt Perfume Museum Foundation.

(5.5) Programme transmitted 26.3.2011. The case was sold for £55.

(5.6) Woolley and Wallis, 21.4.2004, Lot 1633.

(5.7) Ibid. Lot 1100.

(5.8) Ibid. Lot 386.

(5.9) Bonhams 25.11.2004, Lot 52.

(5.10) The firm may have also held the "Harris Patent" for a popular kind of vinaigrette made in Birmingham.

(5.11) By courtesy of Times Archive.

(5.12) See eBay guide to dating perfume bottles.

Fig 6.1

44

CHAPTER 6

SOFT DRINK LABELS

6.1 Many labels, and in particular those of the late eighteenth century and early nineteenth century when England was at war with France, mark the extent and variety of home-made cordials. These drinks were produced in the still rooms and kitchens (see fig. 6.1 for Mrs Smith's kitchen in 1737) of nearly every country house. Patrick Lamb's "Royal-Cookery or The Compleat Court-Cook" published in London in 1716 (6.1) reflects this practice. As early as 1737 Eliza Smith gave detailed recipes for home-made wines and cordials in her "The Compleat Housewife" (fig. 6.2). Elizabeth Raffald in 1799 gave recipes for refining Malt Liquors and curing Acid Wines (fig. 6.3). So did John Farley (fig. 6.4) in 1804 in his "Housekeeper's Complete Assistant" covering home made wines, cordial waters and malt liquors (fig. 6.5). A Bristol blue glass bottle inscribed for "Shallot Wine" has been noted. A hand-written recipe book covering the period 1740 to 1770 contains recipes (6.2) for a range of home made wines including Apricot, Barley, Baum and Birch wines (fig. 6.6). The kitchen was often a long way from the Dining Room and the boudoir, as at Highclere for example.

So the boudoir had to be stocked up with drinks.

6.2 A wide range of fruit juices were produced in the kitchen involving for example apples, apricots, blackberries, black currants, cherries, currants, damsons, elderberries, gooseberries, grapes, grapefruits, greengages, lemons, marmalade oranges, nectarines, oranges, peaches, plums, raisins, raspberries, red currants, strawberries, sultanas, tangerines and white currants. Lemonade sets often comprised a jug with six glasses or beakers. The contents of the jug were identified by a suspended label.

Fig 6.2

Fig 6.3

Fig 6.4

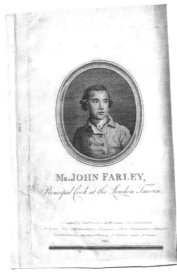

Fig 6.5

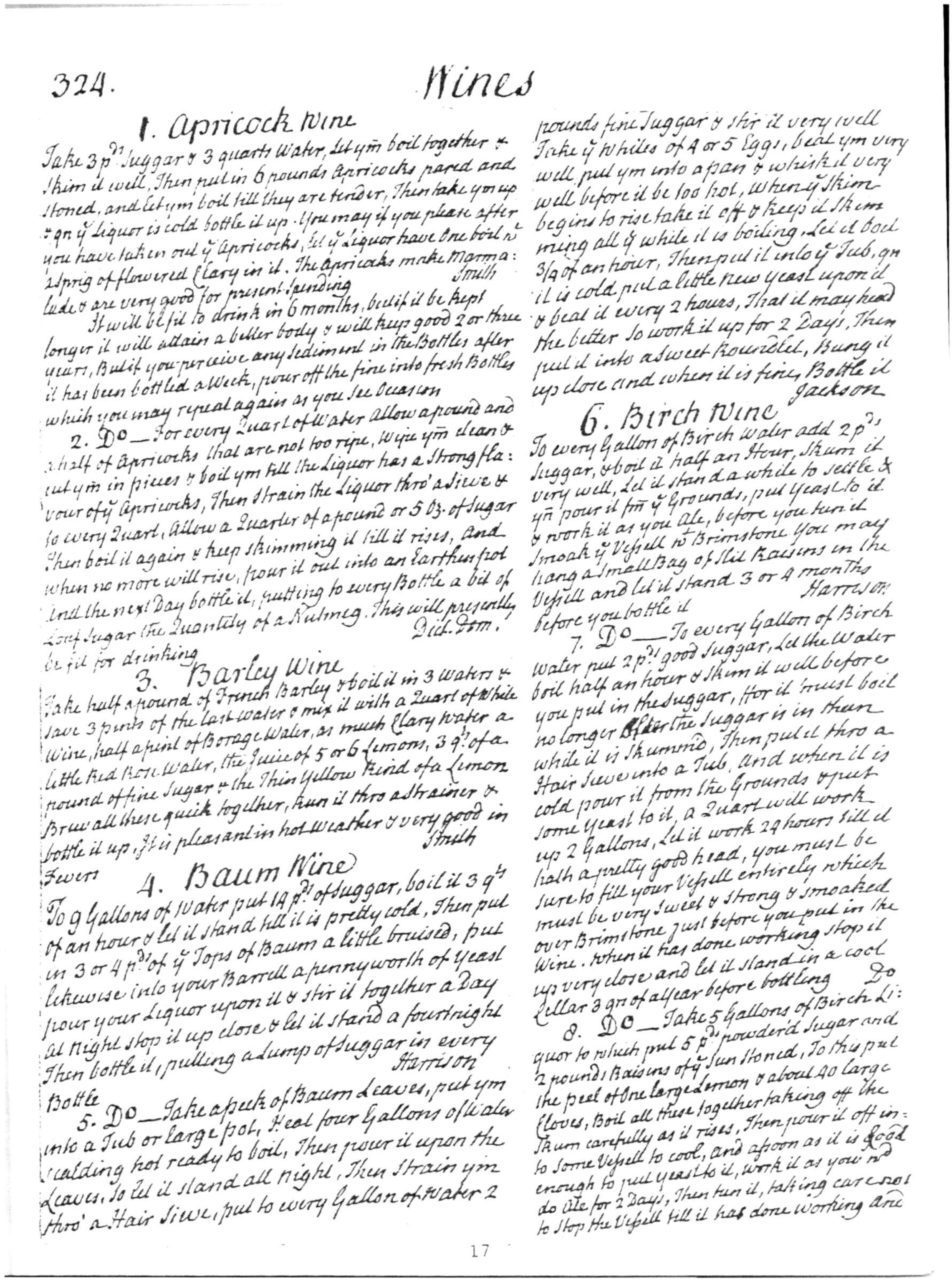
324. Wines

1. Apricock Wine

Take 3 p^d. Suggar & 3 quarts Water, let ym boil together & Skim it well, Then put in 6 pounds Apricocks pared and Stoned, and let ym boil till they are tender, Then take you up ye qn of ye Liquor is cold bottle it up. You may if you please after you have taken out ye Apricocks, let ye Liquor have One boil wth a Sprig of flowered Clary in it. The Apricocks make Marmalade & are very good for present spending. Smith
It will be fit to drink in 6 months, but if it be kept longer it will attain a better body & will keep good 2 or three years, But if you perceive any Sediment in the Bottles after it has been bottled a Week, pour off the fine into fresh Bottles which you may repeat again as you see Occasion

2. Do — For every Quart of Water allow a pound and a half of Apricocks that are not too ripe, wipe ym clean & cut ym in pieces & boil ym till the Liquor has a Strong Fla: vour of ye Apricocks, Then Strain the Liquor thro' a Sieve & to every Quart, allow a Quarter of a pound or 5 Oz. of Sugar Then boil it again & keep skimming it till it rises, and when no more will rise, pour it out into an Earthen pot And the next Day bottle it, putting to every Bottle a bit of Loaf Sugar the Quantity of a Nutmeg. This will presently be fit for drinking Dit. Dom.

3. Barley Wine

Take half a pound of French Barley & boil it in 3 Waters & save 3 pints of the last water & mix it with a Quart of White Wine, half a pint of Borage Water, as much Clary water a little Red Rose Water, the Juice of 5 or 6 Lemons, 3 q^r. of a pound of fine Sugar & the Thin Yellow Rind of a Lemon Brew all these quick together, run it thro a Strainer & bottle it up, It is pleasant in hot Weather & very good in Fevers Smith

4. Baum Wine

To 9 Gallons of Water put 14 p^d. of Suggar, boil it 3 q^rs of an hour & let it stand till it is pretty cold, Then put in 3 or 4 p^d. of ye Tops of Baum a little bruised, put likewise into your Barrell a pennyworth of Yeast pour your Liquor upon it & stir it together a Day at night stop it up close & let it stand a fourtnight Then bottle it, putting a Lump of Suggar in every Bottle Harrison

5. Do — Take a peck of Baum Leaves, put ym into a Tub or large pot, Heat four Gallons of Water scalding hot ready to boil, Then pour it upon the Leaves, So let it stand all night, Then Strain ym thro' a Hair Sieve, put to every Gallon of Water 2 pounds fine Suggar & stir it very well Take ye Whites of 4 or 5 Eggs, beat ym very well, put ym into a pan & whisk it very well before it be too hot, When ye Skim begins to rise take it off & keep it Skimming all ye while it is boiling, Let it boil 3/4 of an hour, Then put it into ye Tub, when it is cold put a little new Yeast upon it & beat it every 2 hours, That it may head the better So work it up for 2 Days, Then put it into a sweet Roundlet, Bung it up close and when it is fine, Bottle it Jackson

6. Birch Wine

To every Gallon of Birch water add 2 p^ds Suggar, & boil it half an Hour, Skim it very well, Let it stand a while to settle & ym pour it fm ye Grounds, put Yeast to it & work it as you Ale, before you tun it Smoak ye Vessell wth Brimstone You may hang a Small Bag of Slit Raisins in the Vessell and let it stand 3 or 4 months before you bottle it Harrison

7. Do — To every Gallon of Birch water put 2 p^d. good Suggar, Let the Water boil half an hour & Skim it well before you put in the Suggar, For it must boil no longer After the Suggar is in then while it is Skimmed, Then put it thro a Hair Sieve into a Tub, And when it is cold pour it from the Grounds & put Some Yeast to it, a Quart will work up 2 Gallons, Let it work 24 hours till it hath a pretty good head, you must be sure to fill your Vessell entirely which must be very Sweet & Strong & smoaked over Brimstone just before you put in the Wine. When it has done working stop it up very close and let it stand in a cool Cellar 3 q^rs of a Year before bottling Do

8. Do — Take 5 Gallons of Birch Li: quor to which put 5 p^d. powder'd Sugar and 2 pounds Raisins of ye Sun Stoned, To this put the peel of One large Lemon & about 40 large Cloves, Boil all these together taking off the Skum carefully as it rises, Then pour it off in: to Some Vessell to cool, and as soon as it is Cool enough to put yeast to it, work it as you do Ale for 2 Days, Then tun it, taking care not to Stop the Vessell till it has done working and

17

Fig 6.6

6.3 Cocoa, coffee, tea and various tisanes were on call from the boudoir. Ginger beer and lemonade were also to hand. There were several different kinds of tea which may have required identification by suspended labelling. Many tea caddies were in pairs or with two compartments, with the addition sometimes of a mixing bowl to achieve a satisfactorily blended tea. However, some tea caddies had compartments for three tea canisters, perhaps for PEPPERMINT, GREEN and INDIAN, for example (6.3). In

MINERAL WATERS,

By *THOMAS DAVIS*,

Purveyor to His Majefty, to His Royal Highnefs the Prince of *Wales*, and the Princefs Dowager of *Wales*, &c.

Are fold at his MINERAL WATER WAREHOUSE, next the St. *Alban's Tavern*, in St. *Alban's-ftreet*, *Pall-mall*, at the following Prices,

	£	s.	d.			£	s.	d.
Seltzer Water, in large Bottles	1	1	0	*Wiltfhire Holt* Water	0	10	0
Bourn Water Ditto	0	18	0	*Tilbury* Alterative Water	0	10	0
The fame in fmall Ditto	0	10	0	*Cheltenham* Water	0	10	0
Pyrmont Water in large Bottles	0	15	0	*Bath* and *Scarborough* Waters	0	9	0
Spa Water in large Flafks	0	14	0	*Briftol* Hot-well Water	0	7	0
The fame in fmall Flafks	0	10	0	*Jeffop's* Well and *Stoke* Water	0	6	0
Haragate Spaw Water	0	12	0	*Sea, Acton* and *Epfom* Waters	0	6	0
Nevil Holt Water, from Dr. *Short*	0	12	0	*Kilburn, Dulwich, Dog and Duck*	0	6	0

Scarborough Salts, 2 s. 6 d. *per* Ounce. *Cheltenham* Salts, 2 s. *per* Ounce. *Acton* Salts, 6 d. *per* Ounce.

N. B. The Foreign Waters are taken up at fuch Times only, when they are in full Vigour and approv'd by the Phyficians both of *Pyrmont* and *Spa*, as will more fully appear by feveral Certificates under their Hands and Seals, as well as thofe from the Magiftrates of *Spa*, who have in the ftrongeft Manner afferted and recommended the Skill, Care, and Long Experience of my AGENT at *Spa*, in taking up the *Poubon* Water in the moft proper Seafon, and at the only fit Times; and fecuring the Flafks after the moft improved Method, moft effectually to retain and preferve their Mineral Spirit. And as feveral Journies have been taken both to *Pyrmont* and *Spa*, to fettle the beft Correfpondence, in order to import the Foreign Waters in their utmoft Perfection, fo I have no Doubt but they will meet with general Approbation.

Note, *Bath, Briftol, Cheltenham, Holt, Jeffop's Well, Sea, Epfom, Acton, Kilburn, Dulwich*, and *Dog and Duck* Waters, come conftantly frefh every Week. *Tilbury* Water fold only at my Warehoufe, and by Mr. *Henry Godwin*, in *Old Bethlem*.

☞ The Purging Water of *Jeffop's* Well, whofe fuperior Strength and uncommon Virtues has been laid before the Royal Society by the Reverend Dr. *Stephen Hales*, approved by the moft eminent Phyficians in *England*, and being come into general Reputation, has occafioned other fpurious Waters to be impofed upon the Publick inftead of the Genuine Water of this excellent Spring. It is thereby become neceffary to caution, whom it may concern, that all Bottles deliver'd at the Well, and at my Warehoufe, are fealed with the Impreffion of *Jeffop's Well*, Three Boars Heads on the Front, and round the Seal, *Thomas Davis, Proprietor*.

Mifs Sheppard

758 — Bought of *Thomas Davis*, in St. *Alban's-ftreet, Pall-mall*.

July 28. 2 Fla. Spa Water — 2 – 4

Rec'd the Contents *Thos Davis*

Fig 6.6a

Fig 6.7

the absence of labelling differentiation could be achieved by having stoppers of different designs. A pair of Bilston/South Staffordshire enamelled glass tea containers, circa 1770, had painted designs of Meissen-style birds and reproduction boudoir labels for GREEN and BOHEA. The two stoppers in this case had identical transfer printed designs (6.4). BOHEA is a black fermented tea.

6.4 The ladies drank substantial amounts of water of various kinds such as Bristol, Buxton, Calamity, Clove, Distilled, Drinking, Eau de Nuit, Iced, Malvern, Mineral (fig. 6.6a), ordinary Water, Seltzer, Soda, Tonic and Vichy. Mrs Smith in her "Compleat Housewife" of 1737 sets out detailed recipes for making the following waters: Barley, Black Cherry, Briony, Centaury, Citron, Cleary, Cock, Doctor Steven's Dropsy, Eye, Fever, Gripe, Hungary, Hysterical, Lady Allen's, Lady Hewet's, Lady Onslow's, Lemon, Lily of the Valley, Lime, Milk, Orange, Palsy, Plague, Rose, Snail, Stitch, Stone, Surfeit (Mr. Denzil Onslow's or King Charles II's), Treacle, Vertigo and Walnut. Bath, Buxton (fig. 6.7) and Tunbridge Wells amongst many other places were famous for their spas. Buxton Water is said to have been produced from 5,000-year-old rainfall that had been forced up through 1,500 metres of bed-rock before emerging at St. Ann's spring (fig. 6.7 gives a view of Buxton's spring-fed baths in 1854). Not all spas were successful like Buxton of Gilbert and Sullivan fame. Mineral waters were exploited by a Company from 1815 at Bowerhill selling Melksham Water, having built a pump-house and accommodation for visitors, but its prosperity decreased from 1822 onwards.

6.5 Boudoir snacks or picnic basket items were made available to go with drinks in squat wide necked four-sided jars, the contents being eaten with the aid of cutlery such as a knife and fork. The picnic however could be a very grand affair. The 6th Duke of Portland's extensive assembly of picnic baskets even included a tent for the ladies. Settings were included to allow stylish dining by up to some forty people (6.5). A 1930s picnic could be decoratively set out in a burr walnut picnic box carried in the boot of a 1934 Bentley, for example, which had a special rack to house it. The picnic basket had a drop down front which revealed an oval tray decorated with marquetry upon which stood glasses, backed by the labelled decanters containing the refreshments. A 1905 picnic was carried in two long rectangular baskets strapped on to the sides at the back of a Renault. The Bentley box which opened outwards would have had a drinks tray surrounded by decanters offering a range of refreshments. A picnic could also be a very low key affair. James Dixon and Sons of Sheffield made a silver sandwich box in 1895 which fitted into one side of a leather travelling case and a mounted silver rectangular shaped glass flask which fitted into the other side (6.6). The flask could have had a suspended label as an aide memoire as to the liquid currently contained in the flask. A Britannia Metal Sheffield plated "huntsman's" sandwich box (marked EP BM) could have come in handy on a picnic. An electro-plated picnic set by Hukin and Heath based on a Christopher Dresser

design of circa 1875 was supplied by Asprey and Co. of London to Edward Preston Jones (evidenced by the initials EPJ on the leather case with a drop down front) around 1910. It contained a kettle, teapot, water jug, tea caddy, milk jug, sugar bowl, burner and stand.

6.6 Examples of food related labels have been reviewed in "Sauce Labels" and include:

BEEF	GAME
BLOATERS	PATE SAVONNEUSE
CAPTAIN WHITE'S	PEAR
CELERIE	PICKLE
CELERY	PRESERVED CHERRIES
EGGS	QUINCE
ESSENCE OF ANCHOVIES	TOAST
FOWL	VEAL AND HAM

6.7 Favourite recipes exist for fruits, preserves, jams, jellies, chutneys, pickles, marmalades, mincemeats and chestnut purees. Some recipes were pirated or "improved". This was a matter of great concern to John Burgess who considered it in 1815 his duty to guard the public against numerous impositions practiced daily. By an advertisement published on 11th January 1815 in the Bury and Norwich Post he requested would-be purchasers of Burgess's Essence of Anchovies to check that the label on the bottle corresponded with the published description (6.7).

6.8 A list of Soft Drink labels is set out below.

APPLE	CRÈME DE CACAO	EAU d'ORANGE
APRICOT	Crème de Thé	EAU DE PLUIE
BLACKBERRY	CURRANT	EAU FROIDE
BLACKCURRANT	CURRANTS	EAU MINERALES BL
BLACK CURRANTS	DAMSON	EAU MINERALES N
BRISTOL WATER	DANDELION	EAUX MINERALES
CACAO	DISTILLED WATER	ELDER
CALAMITY WATER	Drinking Water	ELDERBERRY
CHERRY	EAU	ELDER FLOWER
CITRONSAFT	EAU de MIEL	ELDER.FLOWER
CLOVE WATER	EAU DE NUIT	Elderflower
CORDIAL	Eau de Nuit	ESCALOTTE
CORDIALS	EAU D'OR	ESCHALOTTE
COWSLIP	EAU. D'OR	FRUIT
CRAB APPLE	EAU D'ORANGE	G.BEER

SOFT DRINK LABELS

GINGER
GINGER CORDIAL
GINGER-CORDIAL
GINGERETTE
GOLD WATER
GOOSBERRY
GOOSBERY
GOOSEBERRIE
GOOSEBERRIES
GOOSEBERRY
GOOSEBERY
GOOSERERRY
GOSEBERRY
GRAPE
GRAPE FRUIT
GRAPEFRUIT
GREEN
GREENGAGE
GUJAVA
GULDEWATER
HALLONSAFT
INDIAN
INDIAN TEA
Kersevaser
KIRCCH WASSER
KIRCHEN WASSER
KIRCHEVASER
KIROCHWASSER
KIRCH A WASSER
KIRSCHENWASER
KIRSCHENWASSER
KIRSCHWASSER
KIRSCH-WASSER
KIRSHWASSER
KORSBARSSAFT
LEMON

L. JUICE
LIME JUICE
Limonade
MADEHERE
MALVERN WATER
MARMALADE
MEAD
MEADE
MELOMEL
MILK
MILK PUNCH
MK PUNCH
NECTARINE
NOYEO
NOYEU
ORANGE
Orangeade
Paniagua
PARSLEY
PARSNIP
PEACH
PEPPERMINT
PLUM
P.MINT
P.VINEGAR
QUESTCHE
QUETSCH
QUETSCHE
QUETSH
QUUETCHE
RAISEN
RAISIN
RASBERRY
RASIN
RASPAIL

RASPBERRY
RATAFIA
RATAFIA DE FLORENCE
RATAFIE
RATIFEE
RATAFIER
RATIFIA
RATIFIE
RATTAFIER
R.CURRANT
RED CURRANT
Rp BERRY
R. VINEGAR
SELTER
SELTERS
SELTZER W
SELTZER WATER
SHERBERT
SHERBET
SODA
SODA WATER
S. WATER
STRAWBERRY
SULTANA
TANGERINE
THISTLE JUICE
TISSANNE
TONIC
TWININGS BLENDING ROOM
1706
VICHY
WATER
WATER SOFTNER
W. CURRANT
WHITE CURRANT

NOTES

(6.1) On view at the Harley Gallery in Welbeck Abbey.

(6.2) Bonham's sale no. 19520 held on 19.10.2011, Lot 54 (The Mort and Moira collection of English enamels) attributed to South Staffordshire, circa 1770.

(6.3) Shown on Antiques Roadshow, Sunday, 16.10.2011.

See also paragraph 8.13 on tea caddy labels.

(6.4) See 8 WLCJ 1 at p. 17.

(6.5) See further "Fifty Years of Sport in Scotland" Faber and Faber, 1933.

(6.6) Shown in "Bargain Hunt" on 13.10.2011.

(6.7) See further "Sauce Labels" at p.41, "Anchovy".

Fig 7.1

CHAPTER 7

TOILETRY LABELS

7.1 The ladies liked to look good and aids to beauty were essential requirements for the boudoir. In the eighteenth century , for example, Mr. Keypstick "sold cosmetic washes to the ladies, together with teeth powders, dyeing liquors, prolific elixirs and tinctures to sweeten the breath". So it would not be surprising to find displayed in the boudoir various kinds of creams, lotions, oils, essences, powders, mixtures, salts, tonics, elixirs, washes and soaps. So a dressing-table might become cluttered up with a range of bottles such as those shown in fig. 1.10 and described above in para. 5.20 or shown in fig. 14.1 displaying a carafe of MOUTH WASH, a medicine bottle for BORACIC, a perfume bottle for FRENCH and a soft drink flask for CORDIAL. From 1790 onwards D.R. Harris & Co. Limited as chemists and perfumers have been selling toilet requisites from its Pharmacy in St. James's Street (figs. 7.2 and 7.3). Since 1844 G. Baldwin & Co. have been purveyors of essential oils, herbal tinctures and natural remedies as a leading firm of herbalists, and Taylor of Old Bond Street (now at 74 Jermyn Street) have supplied herbal remedies for hair and skin from Victorian times. From about 1885 the

firm of Charles Horner became well-known for producing gold and silver thimbles and hatpins, well displayed at the British Industries fair of 1920 with hatpin holders.

7.2 In Victorian times Baroness Staffe made the point that some ablutions took place in the Dressing Room. So at Belmont the late Lord Harris' "Blue Dressing Room" contained a washstand with Minton water jug, basin, soap container and tooth-brush holder, whereas in the "Blue Room" next-door (the principal family bedroom) a "characteristic Samuel Wyatt touch is the apsidal alcove with a fitted marble-topped washstand between the windows" (7.1) which is partly screened by the dressing table. Some marble topped washstands were made with cast iron legs circa 1900. The City University Club in London has a late Victorian marble-topped table with a single shelf above for "WET BRUSHES" in its cloakroom. On Lord Harris' washstand stood a Minton water jug, basin, soap container, tooth-brush holder and in his case a pair of matching candlesticks. Robert Heron & Son made hand-painted Wemyss ware sponge bowls incorporating a drainer in its potteries at Sinclairtown, Kircaldy, in Scotland during the period 1920 -1929. A Wemyss porcelain set, comprising a water bucket or slop pail with raffia handle (the lid was missing), two chamber pots or so-called jerries, a beaker, hot water jug and basin, a sponge dish and a soap dish has been noted, retailed by Thomas Goode & Co. of South Audley Street in London. Further Wemyss examples dating from 1882 with the same retailer can be

Fig 7.2

Fig 7.3

seen at Chastleton House near Moreton-in-Marsh, Oxfordshire. With further regard to ablutions a typical Mason's Ironstone set is illustrated (fig. 7.1) without the candlesticks but including a pair of chamber pots or jerries which were to be seen throughout the upper floors of Belmont which were often housed as a pair in self-standing wooden pillars for use as bedside tables known as "night tables" (7.2). A late Victorian set similar to Mason's Ironstone in washing arrangements was made by R. Cox & Co. in Ireland (possibly for Lissanoure Castle) in the "Dresden Sprigs" design, but the soap dish is rectangular and lidded as is the toothbrush and toothpaste holder.

7.3   Water closets are not found above ground level at Belmont. The Mediaeval garderobe or privy closet was replaced by the commode which was supplemented by the bed and douche pan (7.3). The flushing lavatory basin, conceived from early times with the use of water as a cleansing agent, derived from patents granted to Alexander Cumming in 1775 and Joseph Bramah in 1778, was gradually introduced. The invention of the U bend cut down obnoxious smells. Soft paper wipes were introduced by boxed Bromo Paper. The flushing lavatory basin was made famous by Thomas Crapper & Co. established in 1862 with its design comprising a pedestal, wash-down closet, high –level cistern (the so-called patented "Waterfall"), throne seat, pull chain and brass fitments. An authentic twelfth-scale model (fig. 7.4) was made, with the assistance of Thomas Crapper & Co., by Patricia Davis for an International Dolls House Convention in Birmingham. It was actually based on a design of around 1772 by Doctor Erasmus Darwin (1731-1802) which can be seen as an illustration with notes in his design book kept at Lichfield Museum, his former home. He was Charles Darwin's medical grandfather. He belonged to the Lunar Society of Birmingham and was the inventor of the canal barge lift perfected by Anderton. He married into Josiah Wedgwood's family. He did not patent his inventions, perhaps because he was inspired by much earlier studies of the subjects. Another famous original loo used by Charles Dickens from 1856 and accessed by a tunnel has been

Fig 7.4

preserved at Gad's Hill Place. The flushing lavatory basin would of course only work with a water supply. By 1950 some 80% of British homes had flushing arrangements (7.4). Clarence House required renovating before becoming the London Home of the Princess Elizabeth and the Duke of Edinburgh after their wedding in 1947. Although there were six baths in the house, five were in dressing rooms and only one in a room of its own which became known as a bathroom. The bell-board in Stansted House overlooking the coast on the boundary between Hampshire and Sussex shows that Lady Bessborough could summon a maid to a room described as her "BATHROOM". The Duke of Norfolk when modernising Arundel Castle put in 65 flushing lavatories and ten bathrooms.

7.4   Washing and bathing were essential to personal hygiene and afterwards it was customary to apply sweet smelling oils to the skin, taken from displayed bottles decorated with suspended labels. A set of three silver mounted toilet bottles were made by or for Asprey in London in 1956, with engine turned decoration. The bottles were inscribed for PEROXIDE, BATH OIL and FLORIMEL (7.5). A similar but later bottle was untitled. Peroxides have a bleaching effect and are therefore added to some hair preparations. Bath oils afford a luxurious way of pampering oneself with sweet smelling fragrances, such as Moroccan Rose Otto, in moisturising the skin. Florimel, who has featured in a number of myths and fables, was a damsel of great beauty, cast by Proteus into a dungeon but then set free upon the express orders of Neptune. She thus gives her name to sweet smelling products.

Fig 7.5

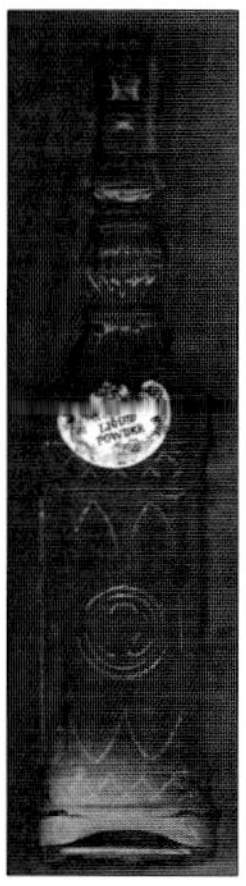

7.5   Dentifrices (there are at least seven variant spellings!) would have been displayed on or near the round table or washstand along with

Fig 7.6

Fig 7.7

by grant of a Royal Charter of incorporation. A beautiful small rectangular silver box contained a mirror set into its lid and a lipstick for carrying out make-up to a high standard. Silver handled brushes of all kinds seem to abound. A ladies silver box contained a small very sharp "Rolls" razor with six spare blades - one for each day of the week!

various sorts of toilet vinegars, toilet lotions, toilet powders and toilet waters. A green glass bottle containing LIQUID POWDER is illustrated in fig. 7.5. A selection of toothpaste powder pots dating from the late Eighteenth Century to the 1930s is illustrated (fig. 9.239a). Mr. W. Woods, a chemist in Plymouth, retailed "Cherry Tooth Paste" in a circular pot during Victorian times (7.6).

7.6 Good grooming was important for men. Wig powder had to be kept in a powder box as arsenic was involved. Napoleon's toilet service was made in Liverpool for use in St. Helena! One particular problem was hair loss. So various preparations were put on the market to help deal with this. Other preparations coloured one's hair to disguise greyness. In 1864 a medical report on hair loss said it was caused by habitual drinking, violence, far too much studying and thoughtfulness, and the pernicious practice of wearing a non-ventilating hat! So hair oil was imported from India for moisturising the scalp. Home-made hair oils were tried out, these being scented with cinnamon, lemon and cloves. Lavender Pomade, contained in a blue green glass bottle, was expensively housed in a silver holder made in Germany around 1890.

7.7 The importance of the toilette is shown by the magnificent silver employed. A toilet service in the collection of the Bank of England includes two candlesticks, one taper, two soap dishes, two ecuelles and a magnificent mirror made in 1694, the year of the foundation of the bank

7.8 From late Victorian times two or three glass toilet bottles stood on silver or glass trays kept on the dressing table. William Gibson and John Langman, plateworkers and owners of The Goldsmiths and Silversmiths Company, produced a number of beautiful silver trays from 1895 onwards. A pair of two-bottle trays made in 1898 has been noted (7.7). The dressing table and shelves for displayed bottles were all important and owners ensured that objects were displayed to best advantage which accounts for the quality of some boudoir labels. Some bottles displaying labels are illustrated in this book. The attention given to quality and detail is perhaps well illustrated by a circular crystal pot made by the Company in 1908 with a hinged silver top which somewhat miraculously had an inset angled adjustable two-way mirror. (7.8).

7.9 The attention given to personal grooming is well illustrated by examining the contents of toilet boxes. For example in 1788 the Duke of Kent (the father of Queen Victoria) took with him (fig. 7.6) an English toilet box (fig. 7.7) containing, amongst other things, four tortoiseshell razors, a shaving brush, a shaving strap, an oval shaving basin, a shaving jug, a stone buffer, a silver toothbrush holder with six detachable brushes and four nail implements as well as no less than seven glass bottles for toiletries (7.9). It could have included ear wax scoops and toothpicks.

7.10 The Duc de Vicence took with him around 1805 a teapot, cup and saucer, an oval shaving basin, a hot water pot with detachable handle,

Fig 7.8

Fig 7.9

Fig 7.10

two razors, a shaving strap, a pen knife, tooth picks, scissors, a seal and sundry other items (fig. 7.8) as well as the silver topped glass bottles for toiletries made by Pierre Leplain of 36 Rue St. Eloi in Paris (7.10).

7.11 Pierre Leplain, along with Marc Jacquard and Francois-Dominique Nodin, made some of the contents of a necessaire made by Hebert around 1810 (fig. 7.9), this time including dining necessities, along with collapsible candlesticks and a lampstand, and forty-seven toilet articles made from various materials including silver, mother-of-pearl and steel. Some were even mounted in gold. There were eleven silver mounted cut glass bottles for toiletries (7.11).

7.12 Another example made around 1830 comes from France but has fittings in silver-gilt with applied armorials for the Ferguson family (fig. 7.10). As well as a tea set with hot water pot and heater there were writing materials, picnic items and extensive toiletry articles including four rectangular glass boxes in three sizes all with silver gilt covers, four faceted glass jars with silver gilt covers, four faceted glass bottles with silver gilt covers, a cylindrical jar, an eye wash and funnel, two ivory brushes and

some seventeen further ivory and metal toilet articles. In 1937 many of the former contents were replaced by English made items such as a stand up mirror, a silver gilt beaker, a medicine spoon, a manicure set, square and rectangular boxes, a match striker, a cylindrical container with a removable shaving brush and a tape measure (7.12).

7.13 Halstaff and Hannaford of 228 Regent street were Victorian case makers housing Gentlemen's travelling dressing table sets. S. Mordan & Co. were often the retailers. Layouts were often similar, with two or three scent bottles capable of taking silver labels, three or two toilet jars and a set of various silver topped glass containers for items of toiletry. Such silver was often made by John How (7.13). A plaque on one toilet box is inscribed "Part of Keel of HMS Hannibal damaged at Kisborn, 17th October 1855". This ship was on active service during the Crimean War under the command of Admiral Sir John Charles Dalrymple-Hay. The silver in this box was made by James Vickery (7.14).

7.14 Mappin & Webb, during the years 1920-1923, produced an art deco 18 ct. gold mounted blond coloured tortoiseshell toilet set (fig. 7.11). It contained a pair of square cut glass toilet bottles. There were however no gold labels. The manner of decoration of the glass is evidence that labelling would have been inappropriate. Other dressing sets of this period contained brushes, a mirror, a nail-

Fig 7.11

buffer or comb but no toilet bottles. A

splendid example was made in London in 1922 of tortoiseshell and silver. A 1903 set included a dusting brush with silver handle. In the same year

TOILETRY LABELS

William Comyns made a magnificent silver Dressing Table mirror with clothes brushes.

7.15 The interest in toiletry labels, as indeed in all boudoir labels, lies not only in their design, shape, appearance, makers and materials used but also in their titles and what these stood for, their history and influence. Within appropriate settings they added great distinction to displays.

7.16 A list of toiletry labels is set out below.

ALCOHOL
ALKOHOL
ALMOND CREAM
AROMATIC VINEGAR
ASTRINGENT
BATH ESSENCE
BATH POWDER
BATH SALTS
B, CURRANT
B.CURRANT
BLACKCURRANT
CLOVE WATER
CLOVES
COLORLESS HAIR TONIC
CRÈME DE FLEURS
    D'ORANGE
CRÈME DE VANILLE
DENTIFICE WATER
DENTIFRICE
DENTIFRICE WATER
DENTRIFICE
DENTRIFICE WATER
DISTILLED WATER
EAU ASTRINGENTE
EAU BORIGUEE
EAU BORIGUEES
EAU BORIGUES
EAU BORIQUEE
EAU de MIEL
Eau de Nil
Eau de Nuit
EAU DE NUIT
EAU DE PLUIE
EAU.DE.PORTUGAL
EAU DE PORTUGAL
EAU DE TOILETTE
EAU DE TOILETTE VERVEINE

EAU DENTIFRICE
EAU DENTRIFICE
ELDER VINEGAR
ELIXIR
ELIXIR DENTIFRICE
ELIXIR DENTRIFICE
ELIXIR DE SPA
EYE
EYE DROPS
EYE LOTION
EYE WASH
Eyes
F. CREAM
French Vinegar
FRENCH VINEGAR
GARGLE
GLICERINE
GLYCERINE
H
HAIR LOTION
HAIR TONIC
HAIR WASH
HAIR WATER
HONEY AND ALMOND
    CREAM
HUILE
HUNGARY
HUNGARY WATER
Infusion d'Homme
LIMOSIN
LIQUID POWDER
LIQUID SOAP
LISTERINE
LYSTERINE
MOUTH WASH
ODC

O.D.C.
ODV
O.D.V.
ODV de Danzic
OEILLET
OIL
PEROXIDE
POMADE
POMAR
POMARD
POMERANS
POMERANZ
PONDS EXTRACT
PORTUGAL
P VINEGAR
ROSE WATER
ROSE.WATER
ROSE-WATER
ROSEWATER
ROSEWATER & GLYCERINE
SPRUCE
TOILET LOTION
TOILET POWDER
TOILET VINEGAR
TOILET WATER
TONIC
TOOTH MIX:
TOOTH MIX 1829
TOOTHMIX
TOOTH MIXTURE
VERBENA
VINAIGRE
Vinaigre Ordinaire
VINEGAR
WHITE VINEGAR
YLANG-YLANG

NOTES

(7.1) Guide to Belmont, Faversham, Kent, p17.

(7.2) Privies of Perfection, Christopher Stevens, Antique Collecting, April 2011, pp 40-45.

(7.3) A splendid example was shown at Cookstown during Antiques Road Trip to Ireland on Monday 17.10.2011.

(7.4) Ibid., Privies of Perfection, p 40.

(7.5) Woolley and Wallis 26.10.2011, Lot 434.

(7.6) Bargain Hunt 26.10.2011.

(7.7) Flog it! 10.10.2011, visiting Melksham in Wiltshire.

(7.8) Flog it! 13.10.2011.

(7.9) Christie's 19.10.2004, Lot 212. See also Chapter Three above.

(7.10) Le Curieux, reference 0041, for sale in October 2004.

(7.11) Christie's 19.10.2004, Lot 168.

(7.12) Christie's 19.10.2004, Lot 156.

(7.13) Christie's 19.10.2004, Lot 803.

(7.14) Christie's 19.10.2004, Lot 805.

Fig 8.1

CHAPTER 8

WRITING INK LABELS

8.1 In Chapter Three, The Travelling Box, mention has been made of writing materials being contained in the Duke of Kent's 1788 box, Edwards' 1825 military campaign box and Neale's 1850 box. Such writing materials may well have included writing implements such as (i) quill pens which lasted until about the 1920s and were widely used by scriveners, solicitors' clerks, writers such as Jane Austen who had her inkstand and quill pen set out on a small table in the cottage at Chawton (in the Austen Museum) during the period 1810 to1815 when she had four books published, and as Thomas Hardy who put the name of the novel he was writing on the relevant writing instrument (in the Hardy Museum), and bank clerks, the Bank of England using over one and a half million quill pens a year at one time (some quill pens are illustrated in fig. 8.1 placed in a Lowestoft china Town Clerk's pot dated 1791) and quill trimmers; (ii) steel nibbed pens which were gradually introduced from the 1830s, often placed on pen rests when not in actual use to catch the drips, and pen wipers (fig. 8.2); (iii) silver pen holders combined with a propelling pencil and owner's seal such as were made by R.M.Mosely & Co. in the 1840s; and (iv) fountain pens which became very popular after the patent was granted in 1884.

8.2 The box or writing slope would also have contained various kinds of writing inks contained in two or three or sometimes more inkwells, being small receptacles of various kinds of shape and design for holding and keeping safe liquid ink, writing parchment or paper, pounce or sand to sprinkle over the paper with fresh writing on it to help dry the ink and prevent it from smearing, paste papers or sealing wax with a seal used to seal letters (sometimes kept in a gold seal case) and a penwipe (fig. 8.2 showing an example

by Sampson Mordan & Co of London in 1900 perhaps made for the Paris Exhibition) as steel nibbed pens had a great tendency to drip and spoil the writing paper. Sealing wax could be dispensed from a wax-jack or from a wax holder with a silver handle which could be dipped into a candle flame or from a candlestick. Other useful additions would have been a calendar, rulers, blotting paper, page markers, stamp box, letter openers, page turners, paper knives (8.1) and combined pen and paper knives.

8.3 The pounce pot or sander would not have required identification and thus labelling. This is because of its distinctive concave rim design (8.2). A fine, slightly abrasive, powder called pounce was put in the pounce pot which had several holes in its lid and the pot was then used to sprinkle pounce over the wet ink to hasten the drying process. The paper or parchment was then slightly folded to enable the pounce to be returned to the pot. The concave rim helped to steer the pounce back into the pot. Pounce was sometimes prepared from sand, salt, or pumice with a slightly chalky quality by mixing various ingredients together. It was also used to size paper and to make paper more receptive to ink. For example it could absorb grease from paper caused by handling. It helped to produce fine, sharp writing.

Fig 8.2

8.4 A writing slope is illustrated (fig. 8.3) and a desk top silver standish is illustrated (fig. 8.4) in set with identical pots needing labels to identify their contents. The labels shown (not in set) are B, probably for BLACK rather than BLUE or BROWN, and COPYING, for use with a printing machine. Another version, not illustrated, has a slope and plenty of storage room, with a stationery pocket in the lid, significantly three compartments to house three square cut glass bottles, a lift out tray with with five racks for pens, various other containers for seals, postage stamps and accessories and a secret drawer, all secured with a Bramah locking device.

8.5 Doctor Johnson records that Alexander Pope (1688-1744) "punctually required that his writing box should be set upon his bed before he rose", presumably so that he could recollect and record his overnight thoughts. Writing materials were clearly an important part of boudoir accessories just like medicines, perfumes, snacks, soft drinks and toiletries. Stationery and postage stamps could be kept available in wooden travelling cases. Parkins and Gotto of 24-25 Oxford Street produced handsome wooden desk organisers from the 1850s (8.3). Some Victorian brass-bound mahogany writing slopes had spaces for three inkwells containing different kinds of ink in bottles (8.4).

8.6 Sir Robert Walpole owned a standish made by Paul de Lamerie in 1733-34 which had three ink bottles, a silver capped pounce container and a silver capped copying ink container. This was gifted to Peter Burrell in 1758 and is now in the Bank of England's Museum. Some ink-stands had accommodation for a bell to summon staff to deal with correspondence, as in the case of the Fruiterers' Company with an example dated 1741, which also had a sander.

8.7 Sir Walter Scott's study at Abbotsford contains writing materials with red and green inks. First book proof revisions were carried out in red ink and second in green ink.

8.8 From around the 1760s the see-into glass inkwells were often replaced with opaque inkwells such as those made in silver or silver-plate, both in the travelling case and on the standish or inkstand. Suspension labels were often needed to distinguish which inkwells contained black, blue, brown, green, red or copying ink. In one campaign chest (illustrated in fig. 3.9) copying ink may have been kept in the larger silver travelling ink-pot with a writing ink being kept in the smaller travelling ink-pot. So sometimes size of inkwell may have been used as a distinguishing feature.

8.9 Brown ink was used for annotating Historia Plantarum in the Sixteenth Century (8.5). Jane Austen made her own inks as evidenced by her

Fig 8.3

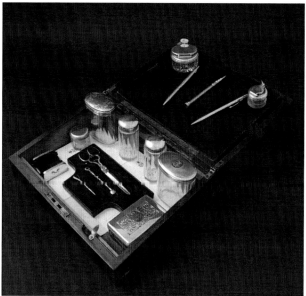

Fig 8.4

recipe "How to make ink" contained in Martha Lloyd's "Household Book" compiled circa 1810 to 1817 full of recipes beloved by the family lodged in the Austen family home at Chawton near Alton. One of the ingredients was gum Arabic. This Household Book was used by and for Jane Austen during the last seven years of her life when she lived there.

8.10   Some inks apparently faded and this was of concern in the development of COPYING ink. The Tenor in the Prologue to Benjamin Britten's "The Turn of the Screw" sings "I have written in faded ink" as he outlines a curious story about to unfold. The "Compleat Housewife" contains a recipe for making ink. The original recipe for Lea and Perrins' Worcester Sauce, which was found in a skip (8.6) by a former accountant at the Company, was written in two different styles of handwriting in sepia ink.

8.11   Some campaign boxes incorporated a printing machine. This comprised a heavy brass cylinder which is turned manually, by the use of a handle, over a sliding surface, which enabled the speedy copying of maps, instructions, despatches and other urgent documents in the field of battle or at a military base (8.7). A special type of copying ink was carried for this purpose. A set of three opaque inkwells has been noted with suspension labels for BLACK, RED and COPYING (8.8).

8.12   Silver labels hung on inkwells on a standish or in an open travelling case would have needed frequent cleaning and it is therefore significant that the silver label for COPYING (see

illustration in Chapter Nine) has a beautiful silver slip chain attached to it. Some inkwells were self standing, as in the case of George Fox's silver mounted hoof made in 1863. In later times novelty ink-wells became popular, such as one shaped like a bell produced by Aspreys in 1909 and a pottery example shaped like a curling stone. These did not accommodate labels.

8.13   It is possible that the small labels for BLACK and GREEN were intended to be used on a tea caddy. A turtleshell box caddy with two compartments marked respectively GREEN and BLACK has been noted (8.9), as well as a container marked BOHEA.

8.14   A possible candidate for a RED label could have been an inkwell at Chatsworth. Convinced that she was about the die, Georgiana Cavendish, Duchess of Devonshire, wrote a message to her only son, then a baby, in the following terms: "As soon as you are old enough to understand this letter it will be given to you. It contains the only present I can make you - my blessing, written in my blood". She would probably have used a scarifier or quill-cutter to obtain a supply as the doctors used leeches for blood letting.

8.14.   The only certain writing ink label is the silver label for COPYING. The list of putative writing ink labels (see fig. 9.33) is as follows:

B	COPYING	SEPIA
BLACK	GREEN	VIOLET
BLUE	PATENT	WRITS
BROWN	RED	YELLOW

NOTES

(8.1)   See further the website of the Writing Equipment Society. Edward Barrett was an expert in desk organisers. Novelty bone combined pen and paper knives were also popular.

(8.2)   An example of a pounce pot in fruitwood can be seen at the Museum of London (A10275); an example in Caughley porcelain can be seen in Shrewsbury Museum (Sy 0857).

(8.3)   See their catalogue for 1856.

(8.4)   Bigwood's auction, 27.5.2011, Lot 497.

(8.5)   See further under MINT in Chapter Nine.

(8.6)   See Daily Telegraph, 3.11. 2009.

(8.7)   See further Antigone Clark's "Antique Boxes 1760-1900" and Antigone Clark and Joseph O'Kelly's "Antique Boxes 1700-1880".

(8.8)   As advised by Schredds of London. All three title are recorded in the 2010 Edition of the Master List.

(8.9)   Shown on "Flog it!" transmitted on Sunday 10.4.2011.

Fig 9.1a From William Henderson's "The Housekeeper's Instructor", W and J Stratford, c1790

CHAPTER 9

COMMENTARY

9.1a   A BRANDY (Medicinal)

Apricot brandy is in set with C BRANDY perhaps for cooking brandy and G BRANDY for ginger brandy. They are large (4.9 x 2.9cm) plated labels with beaded edges in the Marshall Collection, numbered 213, 214 and 217. These could have been drunk in the boudoir with remedial effects, or could have been used in the kitchen for flavouring.

9.1   ACID (Medicinal)

ACID is a mild stimulant. Paper labels exist for at least two acids - benzoic acid and cinnamic acid. A silver example in the MV Brown Collection (1) shows that it could be taken with one in a travelling toiletry case

Fig 9.1

for personal use. Another example of a small silver label for ACID is a star design made by Edward Livingstone of Dundee around 1795. He also made a similar label for RUM which hints at medicinal use for this well known spirit which has been illustrated in the Wine Label Circle Journal (2). ACID is a somewhat general title and one would have to know which particular ACID was in the bottle before using it. An ACID elixir of vitriol was for example used for hypochondriacs and for persons "afflicted with flatulencies arising from relaxation or debility of the stomach and intestines" (3). ACID could also be ordinary vinegar (4) which indeed is itself acidic and used for medicinal purposes (5) and perhaps contained in a bottle in the dressing room labelled with the illustrated beautiful broad crescent label in mother-of-pearl, with border of 12 six pointed stars, of unique design and with a silver chain (fig.9.1).

9.2   Acqua di Parma (Perfume)

This Italian perfume was first produced by hand-distillation in 1916 in a small factory in Parma. It was lighter than most of that time, elegant and

sophisticated. A blend of citrus, rosemary verbena and Bulgarian roses gave it "the smell of Italian summers". So it was "redolent of Mediterranean summers" and it was a "classic zingy scent" suitable for both men and women including Audrey Hepburn, Cary Grant, Ava Gardner and Sharon Stone (6). The original scent was subtitled Colonia and intended to perfume

Fig 9.2

handkerchiefs (fig.9.2). A bottle ticket with this title is plated and of the art deco period.

9.3   ACQUA D'ORO (Perfume)

This oval shaped probably unmarked silver perfume box label was part of a set of eight along with CRÈME de CAFÉ, CRÈME de NOYEAU ROUGE, CITRONELLE, ELIXIR DE SPA, ELIXIR, GRANDE MAISON and HUILE DE VANILLE (7). Notwithstanding this significant grouping it is included in the Wine Label Circle's 2010 Alphabetical Master List (8) as a wine, spirit, liqueur or alcoholic cordial. It is mentioned in Forzano's libretto for Puccini's opera Suor Angelica: "lavoro, quando il getto s'e in dorato, non sorebbe ben portato un secchiello d'acqua d'oro sulla tomba di Bianco Rosa?" This opera had its premiere at the Metropolitan in New York in 1918. Sister Angelica, having resolved on suicide, gathered various herbs and flowers together and made from them a poisonous draught which she drinks and then dies. However A. Rowland & Sons of 20 Hatton Garden in London, perfumiers, placed an advertisement in the Globe and Traveller for Wednesday evening 26 May 1852 to the effect that ladies who visited fashionable races and fetes would do well to provide themselves with a bottle of Rowland's Aqua d'Oro, a fragrant and

WEDNESDAY EVENING MAY 26, 1852.

The Races and Fetes—Ladies who visit these fashionable resorts would do well to provide themselves with a bottle of Rowlands' Aqua d'Oro. This fragrant and Spirituous perfume refreshes and invigorates the system during the heat and dust of summer, and will be found an essential accompaniment for the opera, the public assembly, and the promenade. In all cases of excitement, lassitude, or over-exertion it will prove of great advantage taken as a beverage, diluted with water. Price 3s. 6d. per bottle.

Fig 9.3

spirituous perfume which vitalized, refreshed and invigorated the system during the heat and dust of summer (fig. 9.3).

It could be regarded as an essential accompaniment for the opera, any public assembly and the Victorian or Edwardian promenade. Rowlands, according to an advertisement in the Globe and Traveller, also produced KALYDOR which they hoped "will also prove a most refreshing preparation for the complexion, dispelling the cloud of languor and relaxation, allaying all heat and irritability, and immediately affording the pleasing sensation attending restored elasticity and healthful state of the skin. Freckles, Tan, Spots, Pimples and Discolorations are effectually prevented and eradicated by its application; and in cases of sunburn, stings of insects, or incidental inflammation, its virtues are universally acknowledged".

9.4 ALBA (Perfume)
This title will be found in the Master List (9). It refers to a delicate perfume made from white flowers. It has been mentioned in the Circle's Journal (10). Presumably ALBA is a shortened version of ALBAFLORA.

9.5 ALBAFLOR (Perfume)
Recorded by Penzer in his "Book of the Wine Label" and perhaps for this reason included in the Master List (11), ALBAFLOR is an Old Sheffield Plate label numbered 2 in the Marshall Collection (12). It is of rectangular shape. The title has niello filled lettering. It measures 4.6cm by 2.3cm and so is rather on the large side. This indicates that it was displayed on a large toilet water container. See ALBA-FLORA below for a description of this perfume.

9.6 ALBA FLORA (Perfume)
Recorded by Penzer in his "Book of the Wine Label" and perhaps for this reason included in the Master List (13), ALBA FLORA is a silver label numbered 3 in the Marshall Collection (14) made by Margaret Binley between 1764 and 1778. See ALBA-FLORA (below) for description of this perfume.

9.7 ALBA-FLORA (Perfume)
ALBA-FLORA is a silver label made by Margaret Binley between 1764 and 1778 (illustrated by fig.9.4, currently

on display in the Ashmolean Museum in Oxford, reference WA 1957.24.3.551). It is numbered 551 in the Marshall Collection (15).

Fig 9.4

The label is of domed rectangular escutcheon or cushion shape, the title being pierced, with decorated borders and measuring 5cm by 3.1cm. Alba means white. Flora in Roman mythology was the goddess of flowers, the personification of the power of nature in producing flowers. Thus ALBA-FLORA was a perfume made from white flowers, like WHITE LILAC and WHITE ROSE. It was highly prized as being pure and good for the complexion and very handy to have in one's boudoir. The catalogue of the Marshall Collection notes that this label, displayed amongst a group of sauce labels, was designed "to decorate toilet water containers". The size is larger than Margaret Binley's WHITE PORT, 4.5cm by 3.1cm, displayed alongside in the same frame (number 550) and of similar design. So ALBA-FLORA's container must have been fairly large.

9.8 ALCOHOL (Toiletry)
ALCOHOL was originally an Arabic

name given to the fine metallic powder used by ladies to stain their eyelids (16).

Fig 9.5

The bottle with this label probably contained an essence or spirit obtained by distillation such as a pure spirit of wine. Three quite different designs of French or South Staffordshire enamelled labels are known with this title and are illustrated in " Wine Labels", numbers 1065 and 1066. These enamels with very feminine pink borders had no place on the

Dining Table. The inclusion of this title in the Master List may relate to a label on a glass vessel containing an alcoholic preparation by way of warning. Significantly labels for AMMONIA and LAVENDAR WATER are of similar design (see below for details), and ALCOHOL has been noted in set with ALMOND CREAM, BORACIC ACID, GARGLE and TOILET WATER (see ALMOND CREAM below). ALCOHOL also acted as a mild antiseptic agent. Several examples of small enamel labels for ALCOHOL are illustrated in figs. 9.5 and 9.6.

Fig 9.6

9.9   ALKOHOL (Toiletry)

This name appears on a South Staffordshire enamel. It was in the H.E. Rhodes' Collection. See above ALCOHOL for details of usage.

9.10   ALMOND CREAM (Toiletry)

A boudoir label with this title was hung around a glass jar containing, for example, a home-made skin-care beauty product advocated by Dr. William Scott in "The House Book" published in 1826. The First Section on "The Toilette – Cosmetics and Perfumes" gave advice. One took a pound of oil of almonds and mixed it with four ounces of white wax to give it body. The mixture was then melted and poured into a warm mortar. One pint of ROSE WATER was then added by degrees. ALMOND CREAM, a liquid, was then applied to the skin to make it fair, smooth and white. Almond Paste was a cosmetic for softening the skin and preventing chaps. Dr. Scott's recipe for this is illustrated in fig. 9.7, taken from his "Secrets of Trade" section. Another version by D.R. Harris (established in St. James's Street since 1790) mixes almond oil with lanolin to prevent dryness and delicately fragrances this with rose geranium. Another recipe involved pulverised almonds with

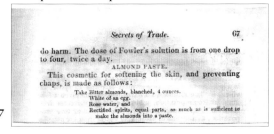

Fig 9.7

natural herbs and spices as additives (see HONEY AND ALMOND CREAM). Like MILK OF ROSES, where rose oil was mixed with beeswax to make a skincare cream, these creams were used in massage therapy. ALMOND CREAM has been found in set with BORACIC ACID, GARGLE, ALCOHOL and TOILET WATER. These labels are made of enamelled copper and decorated with pink roses and green foliage on a white background and would look very attractive in the boudoir.

9.11   AMEIXORIA (Medicinal)

This plum juice non-alcoholic CORDIAL, made from the fruit of the Portuguese plum tree, was thought to be good for aiding digestion after a heavy meal (17).

9.12   AMHONIA (Medicinal)

Standing for Sal Volatile, this was a variant spelling of AMMONIA (see below).

9.13   AMMONIA (Medicinal)

Fig 9.8

This is an antispasmodic antacid, a stimulant like HARTSHORN, and a diaphoretic. It was the chief ingredient of the smelling bottle used to restore swooning Victorian ladies. So it was used to replenish the flacon, a small stoppered transportable bottle. It was also used with gentian to combat gout. An English South Staffordshire enamel with this title is illustrated in "Wine Labels", number 1067, of similar design to ALCOHOL, number 1065. Another enamel label with this title is similar to ALCOHOL number 1066 and is also similar to LAVANDER WATER, having sharp lettering, a pink rose above the title and a garland below it with a yellow flower in the centre. A Crown Staffordshire porcelain label with this title, decorated with a black ribbon, design number A6297, made during the period 1893-1922, was retailed by T. Goode & Co. of South Audley Street in London and is illustrated along with other AMMONIA labels in figs. 9.8 and 9.9.

Fig 9.9

9.14 AMONIA (Medicinal)

This title was used on an American sterling silver label sold at auction (18). It is a variant spelling of AMMONIA (see above).

9.15 ANGELICA (Medicinal)

ANGELICA is a herbal cold remedy, based on a herb in the family Apiaceae valued in New Zealand for its medicinal properties. It is good for respiratory ailments. The title appears on a silver label made in

Fig 9.10

New Zealand by B. Petersen & Co. around 1900, which is marked B.P. & Co. CH CH for Christchurch, South Island (fig. 9.10). It is fairly plain and unpretentious, of elongated shield shape and a pair with C.PORT. Angelica was also used as a flavouring for including in a gin based liqueur and as a flavouring such as Angelica Chateau Yquem for example. C. PORT may have been a restorative used in times of illness or tiredness, or perhaps a cooking port kept handy in the kitchen. These titles are also included in the Master List pending the results of further research into their actual use. ANGELICA was certainly drunk as a herb based liqueur with alleged beneficial properties, and used in the production of Allasch or Doppel-Kimmel or kümel, known later on as simply Kummel. The French version of ANGELICA known as "Angelique" has been mentioned in the Journal (19).

9.16 ANIS (Medicinal)

ANIS, ANISEED, ANISETTE, Anisette blanc, ANNESEED, ANNICETTE, ANNISCETTE, ANNISEED and ANNISETTE are all names included in the Master List presumably referring to the French "pick-me-up" or restorative for use in the earlier morning flavoured with aniseed and therefore called ANISEED with variant spellings. The medicinal properties of aniseed were known to the ancient Egyptians , Greeks and Hebrews. St. Matthew mentions it in Chapter 23 of his Gospel. It was held in high regard as dealing with indigestion. A large plated label in the Marshall Collection (numbered 318) with a significant vine leaves and grapes border pierced for ANNISCETTE measures approximately 6.2 x 5cms. The design and size suggests that this example might not be a boudoir label. Indeed the

Marshall Collection has a glazed pottery bin label (number 436) for ANNISEED. But these names together with ARISETTE were also given to non-alcoholic preparations made from the seed of the anise, which is an umbelliferous plant (bearing umbellate flowers forming umbels or inflorescences borne on pedicals of nearly equal length springing from a common centre) of the genus Pimpinella often confused with dill. ANISEED has the particular quality of helping the human body to expel wind and was therefore used as a medicine. It was sometimes included in the travelling case for use after a good meal. Anise oil is a pure, natural essential oil. ANIS itself is a non-alcoholic CORDIAL flavoured with an aromatic herb in the family Pimpinella known for its liquorice, fennel and tarragon flavour. ANIS appears on a small French enamel mentioned in the Master List, Second Update. See also below HUILLE DE ANIS, HUILE D'ANIS DES INDIES and HUILE D'ANIS ROUGE for undoubted boudoir labels. ANNISETTE appears with CHARTREUSE and COGNAC in a set made in Birmingham by Henry Matthews (1874-1897) in 1880 for the French market, the French import mark of a crab being shown on the face of the labels.

9.17 ANISEED (Medicinal)

This title is engraved upon a small double reeded rounded rectangular silver label (fig.9.11) by Charles Rawlings and William Summers, made in London in 1829 (20), perhaps for

Fig 9.11

use on a small bottle in the travelling box. See above, ANIS, for its particular purpose.

9.18 ANISETTE (Medicinal)

This title appears upon (a) a superb French vase shaped enamel made and marked by Dreyfous, with single roses drooping from the eyelets, with a posy of flowers in the centre, of great femininity (fig. 9.13 and 9.13a); (b) a small French enamel made around 1880 of rounded rectangular format with a triangle top and bottom; (c) a Samson enamel (17); (d) a large escutcheon

Fig 9.12

Fig 9.13

Fig 9.13a

Fig 9.14

shaped enamel with floral decoration numbered 413 in the Marshall collection measuring 5 x 3.5cm; (e) a French copy of a South Staffordshire enamel (fig 9.14);(f) a small elegant boudoir label (21) by Rawlings and Summers circa 1845 (illustrated in fig. 9.12 with neck-ring sauce labels for OUDE and ESCELLOTTE, also made by Rawlings and Summers but in 1832, included for the purpose of size comparison); (g) a small shield shaped silver Dutch label made by Harmanus Lintveld of Amsterdam; and (h) a silver label bearing makers' marks EM and EPC. It is thought appropriate to classify most of these eight different enamel labels as potential boudoir labels. However, ANISETTE was a drink made popular by such personalities as Catherine de Medici, Henry II, Marie Brizard of Bordeaux and Sully. It was said to be composed of spirits of wine, water, sugar, musk, amber, essence of anise and essence of cinnamon (22). Bordeaux therefore is famous for its ANISETTE. This may be due to the success of Marie Brizard and Roger based in the Quartier Saint Pierre as a manufacturer of medicinal cordials from around 1755 and onwards. The commercial liqueur called ANISETTE to-day comes from Bordeaux and Amsterdam, which may account for the existence of Dutch labels. See further ANIS above.

9.19   Anisette blanc (Medicinal)
This title has been recorded in script lettering upon a shield shape French porcelain label. See ANIS and ANISETTE above. Two ANISETTE enamels are illustrated (in fig. 9.13 and fig. 9.14).

9.20   ANNESEED (Medicinal)
This label's title was an alternative spelling of ANISEED. See above, ANIS. ANNESEED was one of the thirteen titles on labelled cordial bottles contained in a Dutch travelling case of around 1750 in date (23).

9.21   ANNICETTE (Medicinal)
This title appears on a plated label in the Marshall Collection (number 318) with pierced lettering and decorated with a border of grapes and vine leaves. It is a very large label measuring 6.2cm by 5.0cm. indicative of substantial usage. It is a pair with SCYRUP. See ANISCETTE above. See ANIS above for commentary on Henry Matthews' ANNISETTE.

9.22   ANNISCETTE (Medicinal)
See ANIS and variants above.

9.23   ANNISEED (Medicinal)
ANNISEED is the name in black lettering on a glazed pottery bin label in the Marshall Collection (number 436) which is indicative of substantial usage and of the need to maintain a supply in store. See further ANIS and variants above.

9.24   ANNISETTE (Medicinal)
See ANIS and variants above. It is in set with CHARTREUSE and COGNAC, both of which were taken for medicinal purposes.

9.25   APPLE (Soft Drink)
Apple juice is a well-known soft drink. Hester Bateman produced a silver object around 1790 perhaps later shaped and engraved for APPLE for use as a label in connection either with the soft drink or with apple sauce.

9.26   APRICOT (Soft Drink)
Small labels probably refer to Apricot juice or to the fruit cordial. APRICOT appears as part of a fruit cordial set made by George Unite in Birmingham in 1910 along with BLACKBERRY, GREENGAGE and RASPBERRY. Another example made of silver in France is in the Cropper Collection at the Victoria and Albert Museum (M 517). APRICOT has been reported in the Journal (24). A recipe for making home-made APRICOCK wine is illustrated in fig. 6.6.

9.27   AQUA MIRABEL (Medicinal)

As indicated by its title, this was a restorative cordial imported from Alsace, made out of small golden - yellow cherry plums, known as Mirabella or Myrobella. It was one of the thirteen titles on labelled cordial bottles contained in a Dutch travelling case of around 1750 in date (25). Eliza Smith made Aqua Mirabilis out of "cubebs, cardamums, galangal, cloves, mace, nutmegs, cinnamon, of each two drams" and then added a range of liquids in a very complicated recipe set out on page 235 of her book.

9.28   ARBAFLOR (Perfume)

Probably an alternative spelling of ALBAFLORA, which see above, this title appears on a curved long double reeded octagonal label by Matthew Linwood, made in Birmingham, in 1812, measuring some 4.5cm. in width (fig. 9.15).

Fig 9.15

9.29   ARISETTE (Medicinal)

This title appears upon a smallish scroll silver label by Charles Rawlings and William Summers, made in London, probably in 1844 (fig. 9.16 showing front and back). The design is elegant, reproducing scrolling baroque. See further above, ANIS. This title is not in the Master List (26). It presumably is an alternative spelling of ANISETTE (see above).

Fig 9.16

9.30   AROMATIC VINEGAR (Medicinal and Toiletry)

Explained in detail in W.B. Dick's "Encyclopaedia of Practical Receipts and Processes", AROMATIC VINEGAR is a compound of strong acetic acid and certain essential oils. One recipe suggests a mixture of one ounce of camphor, one drachm of oil of cloves, and smaller quantities of cedrat, bergamot, thyme, cinnamon and a half-pound of glacial acetic acid. Dr. Scott (fig. 9.17) described Henry's AROMATIC

> **AROMATIC VINEGAR (HENRY'S).**
> This is merely an acetic solution of camphor, oil of cloves, of lavender, and of rosemary. A preparation of this kind may be extemporaneously made, by putting a drachm of the acetate of potass into a phial, with a few drops of some fragrant oil, and twenty drops of sulphuric acid.
>
> THIEVES' VINEGAR, OR MARSEILLES VINEGAR, is a pleasant solution of essential oils and camphor, in vinegar. The Edinburgh Pharmacopœia has given a formula for its preparation, under the title of *"Acetum Aromaticum" (Aromatic Vinegar)*. The repute of this preparation, as a means of preserving health from the attacks of contagious fevers, is said to have arisen from the confession of four thieves, who, during the plague at Marseilles, plundered the dead bodies with perfect security, and who, upon being arrested, stated, on condition of being spared, that the use of *aromatic vinegar* had preserved them from the influence of contagion. On this account, it is sometimes called *"Vinaigre des Quatre Voleurs."* The French codex has a preparation of this kind, consisting of an acetic infusion of various aromatic herbs and camphor, which is termed *"Acetum Aromaticum Alliatum," seu Antisepticum, vulgo "des Quatre Voleurs."*

Fig 9.17

VINEGAR as being "merely an acetic solution of camphor, oil of cloves, of lavender and of rosemary. A preparation of this kind may be extemporaneously made, by putting a drachm of the acetic of potassium into a phial, with a few drops of some fragrant oil, and twenty drops of sulphuric acid". Like smelling salts, it was used as a pungent and refreshing nasal stimulant to deal with faintness. William Buchan, writing in 1785, pointed out that weakness, faintings, vomitings and other hysteric affections were often relieved by vinegar applied to the mouth and nose. He could have also mentioned languor, headaches and problems in seeing properly. It was a stored supply for renewing the impregnation of the sponge kept in a vinaigrette. It was also a travelling box label comprised in a set of four along with VERBENA, EAU DE COLOGNE and MILLE FLUERS engraved upon kidney-shaped labels by Margaret Binley between 1764 and 1778 (27). It is probably the same as TOILET VINEGAR since there were a variety of vinegars in use, such as Vinegars of Litharge, Vinegar of Squills and Vinegar of Roses, and a number of recipes. The making of AROMATIC VINEGAR was overseen in London by the Distillers' Company, founded by Sir Theodore de Mayenne, himself a physician who attended upon King Charles I, whose Charter referred to distilling, rectifying and vinegar making. The set of four by Margaret Binley must be one of the earliest sets of travelling box labels which would be taken with one when going for a weekend stay or to a ball or to any kind of gathering. She was also the maker of a set of five small silver labels for ALE, CYDER, PERRY, R.CONSTANTIA and W.CONSTANTIA which may well have adorned bottles in the boudoir (28). See also below, VINEGAR.

9.31   ARQUEBUSADE (Medicinal)

Paired, according
to Major Gray,
with EAU
de MIEL,
ARQUEBUSADE
was a remedy for
gun shot wounds
and EAU de MIEL
an anti-paralytic
lotion. These
would have been
used to identify
the contents of
flacons in the
Campaign Chest
used during
the course of
the Napoleonic

Fig 9.18

wars and afterwards (29). The name is
said to be derived from "arquebus" meaning an old
fashioned handgun and from "arquebusier" meaning
a soldier equipped with an arquebus. The formula
for ARQUEBUSADE, also called ARQUEBUZADE,
AURQUEBUSADE, EAU D'ARQUEBUSE,
Arquebusade Water, L'Eau d'Arquebuse, Eau
Vulneraire d'Arquebuse and Wound Water, dates
from circa 1480. It seems to have been invented
by monks at the Monastery of Saint Anthony near
Vercours in France. It comprised DISTILLED
WATER from a variety of aromatic plants including
ROSEMARY mille feuille. Wound Water is one of
the titles inscribed on the silver-mounted bottles
in an Augsburg medical chest of around 1630
along with Rose Water and Spirit of Sweet Balm. A
formula for it was developed by pharmacists Fabre
& Bouet in Lausanne, Switzerland, in the Eighteenth
Century. There is an invoice in the archives at
Temple Newsam dated 1764 which reads "to 4 botts
of Arquebusade Water @ 5/- each". This shows
that this product was expensive bearing in mind
that the same archived papers contain accounts for
Port at 1s9d per bottle and for Madeira at 3/- per
bottle. There are references to it being a healing
lotion applied to gunshot wounds, sometimes called
"Theden's Wound Water" after Johann Christian
Anton Theden (1714-1797), a famous German
military surgeon. It was thus used as a general
antiseptic. The formula for it was acquired by Arsad
in 1857 and it was certainly manufactured until 1983.
One formula suggests that it was made up of 7.5

ounces of rectified spirit obtained by the maceration
of plants in the "vulnaires" category, 7.5 ounces of
wine vinegar, 1.5 ounces of dilute sulphuric acid and
2 ounces of Honey Water. William Elliott marked
(by overstamping the original maker's mark) a double
reeded rounded rectangular silver label with this title
made in London in 1818 (fig. 9.18). It adorns a bottle
in a campaign box made in 1824 (see figs. 3.5 to 3.6).
It was even said to have been Sheridan's favourite
tipple when mixed with brandy and eau de Cologne!
No doubt these stimulants were also carried in the
travelling box.

9.32   ARQUEBUZADE (Medicinal)

In 1808 Solomon Hougham made a silver label with
this title in London (30) of kidney or crescent shape
infilled with a cartouche bearing the initials of its
owner "PL" (fig. 9.19, front and back). See above
ARQUEBUSADE.

Fig 9.19

9.33   ASTRINGENT (Toiletry)

Previously known as HUNGARY WATER and also
known as EAU ASTRINGENTE, in Edwardian times
ASTRINGENT was a popular lotion for cleansing
the skin. Used for maintaining a good complexion, it
closed up unwelcome open pores on ladies' faces. It
penetrated pores and toned up the skin. It helped to
prepare the skin for moisturising. It removed excess
oil from the skin. In London in 1908 John Thomas
Heath and John Hartshorne Middleton made a plain
flat kidney shaped silver label (fig. 9.20, front and
back) engraved with this title, niello filled. William
Buchan in 1785 gave a recipe for an astringent
tincture: "digest 2 oz. of gum kino in 1.5 pints of
brandy for 8 days"! Calamine lotion, rose water and
witch hazel are all light astringent tonics.

Fig 9.20

9.34 ATTAR OF ROSES (Perfume)

The celebrated ATTAR OF ROSES, sometimes known as Bulgarian Rose Oil or Turkish Rose Attar or Otto of Roses or Ottar of Roses, is a perfume made out of rose oil, which is a fragrant essential oil obtained from Arabian rose petals won by steam distillation and used mainly in perfume manufacture – see ALBAFLORA, ESPRIT DE ROSE, ESSENCE OF ROSE, MILK OF ROSES, ROSE, ROSE GERANIUM and WHITE ROSE. It was part of a fourteen bottle perfume set, of which eight bottles had silver labels of an unusual splayed hoop neck ring design made around 1800, the other labelled perfumes being BERGAMOT, CITRON, CLOVES, LAVENDAR, MINT, ROSEMARY and WOODBINE. The set was assembled in a shagreen or shark's skin case. Dr William Scott gives an interesting recipe (fig. 9.21) for Otto of Roses taken from a work circa 1826 entitled the "Memoirs of the Rose" in his second section on the Toilette. Otto of Roses is obtained from a complex distillation of Rosa Damascena Corollas. Modern perfume sets may comprise Attar of Roses, English Rose, Rose, Rose Smoke, Balsam Rose and Persian Rose, for example.

> OTTO OF ROSES.
> The following is the recipe for making the celebrated otto, or *attar* of roses, from a work recently published entitled the *Memoirs of the Rose*:—"Take a very large glazed earthen or stone jar, or a large clean wooden cask, fill it with the leaves of the flowers of roses, very well picked, and freed from all seeds and stalks; pour on them as much pure spring water as will cover them, and set the vessel in the sun, in the morning, at sunrise, and let it stand till the evening, then take it into the house for the night. Expose it in this manner for six or seven successive days; and, at the end of the third or fourth day, a number of particles, of a fine, yellow, oily matter will float on the surface, which in two or three days more will gather into a scum, which is the ottar of roses. This is taken up by some cotton tied to the end of a piece of stick, and squeezed, by the finger and thumb, into a small phial, which must be immediately well stopped; and this is repeated for some successive evenings, or while any of this fine essential oil rises to the surface of the water. It is said that a hundred pounds weight of roses will not yield above half an ounce of this precious aroma."

Fig 9.21

9.35 AURQUEBUSADE (Medicinal)

In 1809 John Douglas made a large silver label with this title in London of octagonal shape with double reeded borders (fig. 9.22 , front and back). See above ARQUEBUSADE. The attribution to Douglas (Grimwade 1250) is preferred to that of John Death (Grimwade 1249) because of his reputation as a maker of travelling box fittings.

Fig 9.22

9.36 B (Writing Ink)

Rawlings and Summers made a writing ink label to identify Black Ink from any other colour or from COPYING ink. A small double reeded rectangular silver label with rounded ends and a thin delicate short chain, engraved "B", is fully marked for 1840 and is illustrated (fig. 9.23). William Welch of Exeter made upended oval labels engraved for "R" and "B", but their size suggests use for RUM and BRANDY decanters rather than somewhat large ink bottles.

Fig 9.23

9.37 BALM (Medicinal)

A recipe is given by Dr. William Scott (1826) under the heading "The Toilette" in the following terms: "SPIRIT OF BALM. Tops of balm, one pound to the gallon proof", see below fig. 9.87. He also gave a recipe for Solomon's BALM of Gilead (see fig. 9.24). Langtry BALM was an aid to beauty popular in the 1920s. BALM is an aromatic substance, exuding naturally from various trees of the genus Balsamodendron. It was used as an ointment for healing wounds. William Bateman I made a silver label in 1822 of plain reeded octagonal design engraved with this title (fig. 9.24a, front and back). A bin label with this title has been included in the Master List. Spirit of Sweet Balm is one of the labels on silver bottles in the Augsburg set of 1630. The

> SOLOMON'S BALM OF GILEAD,
> An aromatic tincture, of which cardamoms form a leading ingredient, made with brandy. It is also supposed that the Spanish fly enters its composition.
> *.* This boasted nostrum is nothing more or less than a simply medicated dram. The Balm of Gilead is now the slang phrase for the gin-bottle; and every old woman who can afford to purchase a bottle of Solomon, the better to disguise her propensity, extols it to the skies as a sovereign remedy for the "colic and the phthisic;" and in those circles where the Balm of Gilead is not to be attained, a glass or two of Hodges' cordial gin goes down as an excellent substitute. Among modern empirics, Solomon bears the bell. He knew brandy to be the *elixir* best calculated to give energy to the drooping spirits; and also, that, by adding a little of the powder, or rather the tincture, of the Spanish, that a miracle for the moment might be wrought, that would require, every time, an additional stimulus, to produce similar effects; and so on in succession, until the Balm of Gilead becomes as inert as a glass of pump water.

Fig 9.24

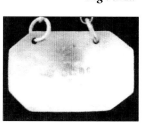

Fig 9.24a

Balm of Mecca is another well-known mixture, although in 1717 Lady Mary Wortley Montague discovered that it made her face swell and her skin turn red!   She should have taken the advice given by the author, whose initials were H.N., of "The Ladies Dictionary : Being a General Entertainment for the Fair Sex" published in 1694 – "A painted face is enough to destroy the reputation of her that uses it". However, as one of the ingredients of this Balm was turpentine, perhaps the outcome was not altogether surprising. No wonder it was said that the party's face always seemed to carry Lent in it. Confusingly BALM was sometimes the name given to the white liqueur called CHARTREUSE made at the Carthusian Monastery near Grenoble, because of its aroma and medicinal properties, which is why CHARTREUSE is found in set with COGNAC and ANNISETTE.

9.38   BARLEY WATER (Medicinal)
A recipe (illustrated in fig. 9.25) suggests that one should boil a quarter of a pound of pearl barley in two quarts of water. You then skim it very clean and when it has boiled half away, strain it. One then puts in two spoonfuls of white wine. It must be made a little warm before you drink it. Another recipe suggests a mixture of barley, sugar and lemon-peel. Eliza Smith on page 221 of her book recommends adding "a blade or two of mace or a stick of cinnamon". The Romans were said to have added white wine and honey to this mixture. Thus it is a demulcent drink, soothing, lenitive, mollifying, designed to allay irritation. It was used to succour the young and the infirm. A recipe for making home-made BARLEY WINE can be found in fig. 6.6.

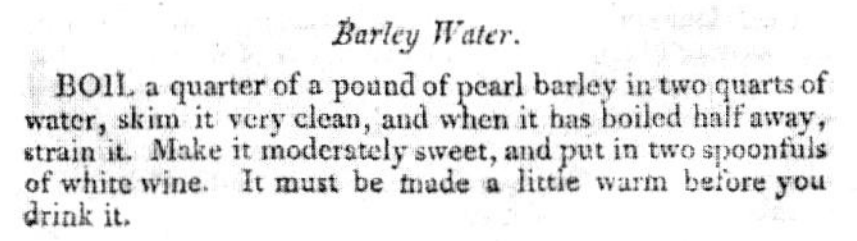

Barley Water.

BOIL a quarter of a pound of pearl barley in two quarts of water, skim it very clean, and when it has boiled half away, strain it. Make it moderately sweet, and put in two spoonfuls of white wine. It must be made a little warm before you drink it.

Fig 9.25

9.39   BATH ESSENCE (Toiletry)
An early 20th Century enamel, decorated with pink roses (fig.9.26), with a transfer printed title of BATH ESSENCE, is similar to BATH SALTS. Presumably it adorned a bottle on a shelf in the washing area

containing some suitable essential oil to be added to the bath water for the purposes of aromatherapy.

Fig 9.26

9.40   BATH POWDER (Toiletry)
This title has been noted on both enamels and porcelain. It appears on a slightly larger label than the norm, of considerable quality and made of enamel in Staffordshire around 1880 with contemporary chain (fig. 9.27). The reverse is creamy white. It also appears in Crown Staffordshire England porcelain dating between 1892 and 1923 retailed by Thomas Goode & Co. of South Audley Street London (fig. 9.27a). Painted in bright colours with a floral design, and originally in the Horlick Collection, the label has a registered design number of A6297 which is also found on LIQUID POWDER, BORACIC LOTION and GLYCERINE. This design number has also been noted on EAU DE ROSE and EAU DE COLOGNE, probably a pair because they have similar chains and lettering. LIQUID POWDER has sloping letters and a small link chain. BATH POWDER has even lettering and a rather rough looking chain. It has been suggested that this label adorned a bottle in the bathroom containing powders to be added to the bath water for the purposes of aromatherapy, but in fact it was a beauty aid rather like talcum powder made out of powdered arries root mixed with an essential oil. It adorned a bottle which stood on the dressing table.

Fig 9.27

Fig 9.27a

9.41   BATH SALTS (Toiletry)
Salts were used for the purposes of aromatherapy.

This title appears on enamel labels of various sizes to accommodate various sizes of bottles used in the bathroom or washing area. Two enamel labels are illustrated in fig. 9.28. BATH SALTS could belong to the same set as BATH ESSENCE, TOILET

Fig 9.28

POWDER, GLICERINE and EAU DE VIOLETTE, all having similar designs, colouring and chains. It is also a title on a small enamel label c. 1890 formerly in the Horlick Collection. It appears in Major Gray's list (31) supplementing Penzer's 1947 list. Salts were often combined with essential oils and herbs. Vanilla extract was often used in making BATH SALTS.

9.42 B, CURRANT (Toiletry)
Standing for BLACK CURRANT this was an energizing oil produced to help ladies maintain a good complexion in all weathers. It was said to boost metabolic activity enhancing purification of the skin. It was infused with lemon, cypress, sage and tea tree oil. It was contained in a large jar with an electroplate label in the Marshall Collection (number 10) sized 4.6 x 2.6 cm.

9.43 B.CURRANT (Toiletry)
See B, CURRANT above

9.44 Benedict (Medicinal)
This title in script format appears on a silver octagonal double reeded label made by John Rich in 1802, probably for use in the travelling box as Benedict was a mild laxative. It has been included in the Master List presumably as it was thought to relate to Benedictine, but Benedictine was not on the market until around 1882!

9.45 BENZIN (Medicinal)
Benzoic acid or BENZIN was a mild stimulant imported from Sumatra and Java by the Dutch East India Company, also used as an antiseptic. It was made from the herb styrax benzoin which gave its name to BENZOIN, BENZOIN LOTION, BENZOIN SALTS and BENZOIN SOLUTION. It has been noted as possibly a pair

Fig 9.29

with ALKOHOL. An apothecary's rounded glass bottle or jar is labelled "AC. BENZOIC". The label is made of paper, of rectangular shape rather like a suspended label, but with fancy gold decoration (fig.9.29). BENZIN should not be confused with benzene, a volatile flammable distillate. Benzoic Acid is used as a food additive for animal nutrition.

9.46 BENZOIN (Medicinal and Perfume)
This title appears on a pottery label. Prada of Milan when considering timeless elegance of fragrances refer to mystic BENZOIN incense. Tincture of Benzoin was used in making MILK OF ROSES. See above, BENZIN.

9.47 BENZOIN LOTION (Medicinal and Perfume)
This title appears on a French Samson factory enamel label of around 1890 (fig. 9.30). See above, BENZIN and BENZOIN

Fig 9.30

9.48 BENZOIN SALTS (Medicinal and Perfume)
This title appears on an enamel label with pink roses. See above, BENZIN and BENZOIN.

9.49 BENZOIN SOLUTION (Medicinal and Perfume)
This title has been noted. See above, BENZIN and BENZOIN.

9.50 BERGAMOT (Perfume)
BERGAMOT is an aromatic essential oil obtained from the rind of the Citrus bergamia, a lemon coloured fruit about three inches in diameter with a smooth skin; the tree is cultivated in Southern Calabria and should not be confused with the variety of pear of the same name, although it is said to have been first produced by grafting a citron on the stock of a bergamot pear-tree. Nor should it be confused with the herb bergamot so-called because it had a similar flavour to that of the essential oil. The herb was used as a flavouring in cooking as well as an infusion for beverages and indeed early American colonists are said to have used bergamot as a substitute for the tea they were boycotting! It was thought to be good for relieving stress and for dealing with acne, abscesses, boils, cold sores and

skin complaints. Essence of Bergamot is, according to Dr. Scott, a constituent of MILK OF ROSES. One glance at the beautiful label in the Cropper collection, a pair with FRANGIPANNI, unmarked but importantly made in silver-gilt, establishes it as a perfume label. Furthermore BERGAMOT was one of a set of eight silver splayed hoop neck ring labels keeping company with ATTAR OF ROSES and other perfumes (see ATTAR OF ROSES above). BERGAMOT and FRANGIPANNI may have been paper titles for the elegant slot labels made in Paris by Armand Gross around 1895 in silver of .950 fineness (fig. 9.31 to the right). The design of the slot label is similar to an unmarked example of BERGAMOT in the Cropper collection (fig. 9.31 to the left). Essence of Bergamot was an essential ingredient in making Nonpareil Water or Eau Sans Pareille said to be a fragrant cosmetic (see below fig. 9.87).

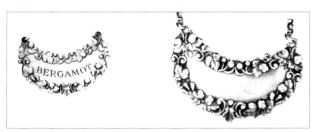

Fig 9.31

9.51 BICARBONATE OF SODA (Medicinal)

BICARBONATE OF SODA, Sodium Bicarbonate or Baking Soda, best known in the culinary arts as a baking powder being a rising agent used in cooking cakes and for softening cabbage, was in this context used for soothing ailments and for helping pregnant women deal with morning sickness. It was also used as a cleaning agent for use in the bathroom, for cleaning out refrigerators and for many other uses in the household. It is very slightly abrasive. It is included in tooth paste to make teeth whiter. BICARBONATE OF SODA was included in an undoubted medical travelling box of around 1850 along with BORACIC ACID, LAUDANUM, MAGNESIUM CITRATE and SAL VOLATILE.

9.52 BICARB SODA (Medicinal)

This title is an abbreviation, see above BICARBONATE OF SODA.

9.53 BICARD DE SOUDE (Medicinal)

This title is the French version of BICARBONATE OF SODA, which see above.

9.53a BITTER (Medicinal)

This title appears upon an oval French ceramic label (fig. 9.31a) and upon a small shield-shaped French label (fig. 9.31b) It is thought that these two labels refer to BITTERS, on which see below.

Fig 9.31a

9.54 BITTERS (Medicinal)

A delicate scroll-bordered silver label pierced for BITTERS is decorated with matted leaves at intervals. It was made by William Evans in London in 1876 (fig. 9.32). The root of gentian is the fundamental "bitter" in BITTERS which refers to bitter medicines generally, such as quinine. So quinine is impregnated with the extract of

Fig 9.31b

Fig 9.32

gentian with quassia and orange peel added, for example, for use as a stomachic. Quinine is a tonic. It is an alkaloid found in the bark of species of cinchona and remigia. So BITTERS were used as a cure for stomach illness. In 1824 in Venezuela under the direction of Dr. Johann Siegert BITTERS were traded from Angostura as a cure for sea-sickness. BITTERS were also known for being an alcoholic beverage, flavoured with herbal essences with a bitter taste, marketed in the 1870s and 1880s according to contemporary advertisements as a patent medicine to cure general disabilities such as liver complaints, sick headaches, biliousness, indigestion, loss of appetite, colds, fevers, coughs, palpitations, jaundice, salt rheum, constipation, dyspepsia, humours, diarrhea and colic. BITTERS were sometimes kept in a chemist's shop in a colourful display corked bottle along with MINT. In the Twentieth Century BITTERS became the essential ingredient of a number of cocktails, such as Pink Gin, and it was used in aperitifs and in digestives. Recipes for CORDIALS, SURFEIT WATERS and BITTERS were contained in a household hand-written book compiled around 1740-1770. The title

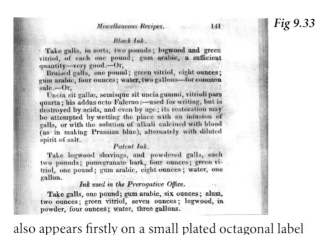

Fig 9.33

also appears firstly on a small plated octagonal label circa 1880 with beaded and fluted edges, in set with R.VINEGAR (for red vinegar), measuring 3.1cm by 2.0cm; secondly on a small kidney shaped label by Paulo Rosso of Malta with maker's mark and silver standard mark only, circa 1860, presumably commissioned by the Royal Navy and illustrated in the Journal (32); and thirdly on a French enamelled escutcheon, where in the title there is a dot over the capital I in BITTERS.

9.55 BLACK (Writing Ink)
Mentioned in the Master List, this title may refer to a silver label hung around an inkwell to identify its contents and distinguish it from other inks of similar appearance such as COPYING ink. Dr. Scott gives recipes for making inks such as BLACK ink, PATENT ink, and WRITS ink used in the Prerogative Office, all under the heading of Miscellaneous Recipes (fig. 9.33).

9.56 BLACKBERRY (Soft Drink)
BLACKBERRY is a fruit juice cordial in set with APRICOT (which see above) and others.

9.57 BLACKCURRANT (Soft Drink and Toiletry)
BLACKCURRANT is a fruit juice cordial in set with APRICOT (which see above) and others. It was also used as a toiletry item. According to Ole Hendrikson it is an energising oil for the complexion which after application forms a protective shield against the elements. Infused with lemon, cypress, sage and tea

Fig 9.34

tree oil it boosts metabolic activity and enhances purification of the skin. BLACKCURRANT could also be a tea caddy label because, along with ginseng and vanilla, it is a fresh and fruity tea (fig. 9.34). Blackcurrants were stored since Wedgwood produced a bin label for BLACK CURRANTS, mentioned in the Master List, First Update.

9.58 BLACK CURRANTS (Soft Drink)
Black currants were stored, as evidenced by a bin label with this title. See above, BLACKCURRANT.

9.59 BORACIC (Medicinal)
BORACIC is like, pertaining to, or derived from borax, a whitish odourless crystal or powder being the acid borate of sodium or biborate of soda also known as BORACIC ACID, ortoboric acid,

Fig 9.35

borofax, three elephant, hydrogen orthoborate and sassolite. Whilst it was used in making cosmetics, it was a popular fungal foot soak and the bottle it marked contained the crystals or powder for adding to the warm water in the foot bath. The title appears on an enamel label with pink borders and the usual pink rose above and floral display below the name (fig. 9.35).

9.60 BORACIC ACID (Medicinal)
See above BORACIC. The title has been noted on a Crown Staffordshire porcelain label of around 1920, and in set with ALMOND CREAM (which see above), ALCOHOL, GARGLE and TOILET WATER. BORACIC ACID, a paper label, was included in the Belfast Medical Box.

9.61 BORACIC LOTION (Medicinal)
See above BORACIC. The title appears on a Crown Staffordshire England label retailed by Thomas Goode &Co. of South Audley Street pattern number A 6297, perhaps a pair with GLYCERINE, both having the painter's mark "F". A solution of BORACIC was used as an eye lotion or eye wash. See fig. 9.36.

Fig 9.36

9.62   BORACIC WATER (Medicinal)
See above BORACIC. This was probably the same
as the lotion. It appears on a plated label of perhaps
post-world-war I date probably in a set with PONDS
EXTRACT, GLYCO-THYMOLINE, LYSTERINE
and CITRATE MAGNESIA.

9.63   BORIC (Medicinal)
See above BORACIC. The title appears
upon an enamel label of medium size.
It is mentioned in Elizabeth Bennion's
"Antique Medical Instruments". An
apothecary's jar in the shop in Blist's Hill,
Ironbridge, is labelled "PV. AC. BORIC"
on a rectangular paper adhesive label.

Fig 9.37

9.64   BORIC ACID (Medicinal)
See above BORACIC. This title appears on an
enamel painted on the reverse in red "France",
formerly in the Lank Collection . The colour of the
border is reddish brown rather than pink (fig. 9.38).

Fig 9.38

9.65   BOUQUET (Perfume)
Dr. William Scott had a recipe published in 1826
for Eau De BOUQUET sometimes called Nosegay
Water (see below fig. 9.87). It involved mainly
odoriferous honey water and nonpareil water with
the addition of a variety of spirits, including jasmine,
cloves, violets, flags, lavender and orange flowers.
From 1828 Guerlain (see below) sold ESSENCE
BOUQUET. From about the 1870s Penhaligans sold
BOUQUET as one of sixteen original fragrances
allegedly based on clean, natural ingredients. The
Household Cyclopedia of Perfumery gives a formula
for Esterhazy Bouquet and a formula for Essence of
Bouquet. Guerlin retailed Bouquet de l'Imperatrice,
Bouquet Napoleon and Bouquet de Furstenberg from
around 1863. In the 1930s ESS. BOUQUET formed
part of a perfume set, along with LAVENDER and
OPPOPONAX, comprising tiny leaf shaped silver
labels with art deco lettering. See also BOUQUET
DE ROI and HAMMAM BOUQUET below.

9.66   BOUQUET DE ROI (Perfume)
This perfume appears to have been invented by
Pierre-Francois-Pascal Guerlain (who died in 1864)
and first sold in 1828 from his shop on the ground
floor of the Hotel Meurice on the Rue de Rivoli
in Paris as an English fragrance, perhaps based
on BOUQUET as he had studied the distilling of
essences in London and named the perfume after
King George IV who is therefore the ROI. Guerlin
in fact called it "Bouquet de Roi d'Angleterre" so that
there could be no mistake. A recipe for Bouquet
Royale is given in The Household Cyclopedia of
Perfumery.

9.67   BOUQUET DU ROI (Perfume)
See above BOUQUET DE ROI.

9.68   BRANDY (Medicinal)
BRANDY was often taken at night for medicinal
purposes. It was a popular drink enjoyed by guests
well after dinner and often taken on the way to
retiring for the night and in the bedroom. It was
used as a tonic for an upset stomach and for this
reason was sometimes carried in the travelling
box. The ladies had a recipe for orange flower
brandy described as a CORDIAL. Boudoir labels
are distinguishable from wine labels on
grounds of size and
significance, although
some of these tiny labels
may have been used on
a spirit decanter kept
on the landing or on
a spirit decanter in a
spirit decanter frame in
a withdrawing room or
on the soy frame in the
dining room as well
as on small decanters
in the bedroom or
dressing room. So
these small labels
fall into each of the
three categories:
wine/spirit, sauce and
boudoir. Rawlings
and Summers
made a tiny plain
reeded octagonal
label engraved
for BRANDY in

Fig 9.39

Fig 9.40

Fig 9.41

Fig 9.42

Fig 9.43

Fig 9.44

1860. A silversmith using the mark possibly AJ (punch rubbed), perhaps for Alexander Johnston, in 1799 made a small double reeded rounded end rectangular with a niello infilled title for BRANDY. The Army and Navy Co-operative Society Limited in 1914 made a reeded small octagonal label for BRANDY. John Reily produced around 1815 a very small label for Cognac. A pair of enamel labels for BRANDY and ELDER FLOWER both fluoresce, so BRANDY kept company with ELDER FLOWER (see below), and they indeed seem to keep company with ALCOHOL, AMMONIA and PEROXIDE. In 1798 Peter and Ann Bateman made a tiny long rectangular BRANDY which was suitable for use in a travelling box. Illustrated are a silver unmarked double-reeded kidney shape (fig.9.39), a George Pearson London marked rounded rectangular gadrooned circa 1817 (fig. 9.40), an elegant scroll plaque designed by Nathaniel Mills and made in Birmingham in 1845 (fig. 9.41), a curved domed feather edge early label attributed to John Bridge (maker's mark fig. 9.43) in the style of Richard and Margaret Binley, circa 1830 (fig. 9.42) and a cast floral basket design by Rawlings and Summers made in 1831 (fig. 9.44) and three small silver labels (fig. 9.44a).

Fig 9.44a

9.69 BRISTOL WATER (Soft Drink)

The firm of Phipps, Robinson and Phipps in 1811 made a plain reeded rectangular label with rounded ends. Near Bristol a hot-spring arose from the bank of the River Avon at the foot of Saint Vincent's rock in the Avon Gorge. The spring water contained a large variety of mineral elements. One analysis carried out indicated the presence of carbonic acid, nitrogen, magnesium chloride, magnesia nitrate, sodium chloride, sodium sulphate, magnesium sulphate, carbonates of lime, bitumen and silica. In 1980 the Bristol Civic Society together with the Clifton and Hotwell Improvement Society put up a plaque about the famous Hot Well which had been praised for its health giving propensities since the Fifteenth Century. In the Eighteenth Century Assembly Rooms were built over and incorporating the well in a Pump Room, which had a colonnade added to it in 1786 and shops were built around incorporating a lending library. According to Matthews in his Bristol Guide nobility and gentry from all around came to "enliven and embellish the town, introduce propriety of diction, taste for literature, novelty or fashion and elegance of address". As at Bath people came to drink the spa water. This led to the water being bottled and sold as BRISTOL WATER or Bristol Hot-well Water, for example by Thomas Davis, purveyor of mineral waters to King George II, at 7 shillings per bottle in 1758.

Fig 9.45

9.69a C BRANDY (Medicinal)

See below C PORT for explanation.

9.69b C; BRANDY (Medicinal)

See below C PORT for explanation.

9.70 C PORT (Medicinal)

This intriguing title appears on a silver label made in New Zealand by B. Petersen & Co. around 1900. It is fairly plain and unpretentious, of elongated shield shape and a pair with ANGELICA (see above). It may have been a restorative used in times of illness or to help in culinary matters as a cooking port kept handy in the kitchen. These

Fig 9.46

titles are also included in the Master List pending the results of further research into their actual use. C. PORT may refer to some remedial aspect of cooking port, possibly as a flavouring for some medicinal products. Cooking Brandy, for example, has been found in set with Ginger Wine and Damson Cordial, and is recorded as C BRANDY, numbered 214 in the Marshall Collection, and as C;BRANDY by Joseph Gibson of Cork circa 1790. Cooking Port was used in great quantities as there is a bin label with this title made by Macord and Arch of 47 Great Tower Street in London. Another bin label at Doddington Hall, Grantham, has the unusual title of Baroness Port. The existence of Cooking Sherry is recorded in the Journal.

9.71　C. De ROSE (Perfume)
A silver plated rectangular label with this title, having cut corners and beaded edges, is numbered 215 in the Marshall Collection, measuring 4.8cm by 2.9cm. It stands for Crème de Rose, a perfume based on rose oil. See further CRÈME DE ROSE below and ATTAR OF ROSES above.

9.72　C. DE ROSE (Perfume)
See above C. De ROSE.

9.73　CACAO (Soft Drink)
A French porcelain label marked CACAO was hung around a jar containing a powder, made from the seeds produced by cacao trees, known as cocoa, which with added hot water makes a popular evening beverage. CACAO is also the name given to a commercially produced liqueur.

9.74　CALAMITY WATER (Soft Drink)
This silver label in the V & A collection is a broad rectangular in shape with a splendid leaf and tendril border. It could well have been used in the withdrawing room designed to help relieve deep distress arising from some adverse circumstance. It may even have been on show in the boudoir.

9.75　CAMPH DROPS (Medicinal)
This presumably refers to drops of CAMPHORATED SPIRITS OF WINE (see below). The title was engraved on a double reeded, rounded ends rectangular silver label made by Charles Reily and George Storer around 1830. It bears a maker's mark only and matches in size and design a set of six labels which includes CAMPHTD SPIRIT. So there could

be seven or more labels in this set. Camphor is a whitish translucent crystalline volatile substance, a vegetable essential oil distilled from Camphora Officinarum purified by sublimation. Dr. William Buchan writing in 1785 explained that camphor is first of all dissolved in alcohol. A pint of rectified spirits is made. Then one puts drops onto a cube of sugar. This is then taken by a patient for the relief of a cold. It was also used as an embrocation, for example for bruises, palsies and chronic rheumatism. It was also used for preventing gangrenes.

9.76　CAMPHtd. SPIRIT (Medicinal)
The distilled spirit has a bitter aromatic taste and characteristic smell. The label title had to be abbreviated to fit on a small label for use in the travelling box. This title has been found on an unmarked silver label numbered 344 in the Marshall Collection in company with EAU D'ALIBURGH, ESSENCE OF GINGER, POISON LAUDANUM, SPIRIT LAVENDER and TOOTH MIXTURE all unmarked but probably made by Reily and Storer around 1830. This attribution is based on the matching marked label for CAMPH DROPS having been acquired by Mrs Marshall according to her notes along with the set at the same time. See above CAMPH DROPS.

9.77　CAMPHOR JUICE (Medicinal)
Camphor oil is a pure, natural essential oil. See above CAMPH DROPS

9.78　CAMPHOR JULEP (Medicinal)
There seem to be two types of Juleps: one based on the Persian gulab or rose water being a sweet drink variously prepared and sweetened with syrup or sugar; the other popular in America from at least 1804, being a mixture of brandy and other spirits sweetened with sugar with added ice and flavouring, especially mint as in Atlanta. The idea was to cool one down, and so it was a popular choice for the travelling box. Thus the title has been found on a small silver gilt oblong label. Thus designed to calm the nerves, CAMPHOR JULEP was made famous by Fanny Burney whom Jane Austen called "our first woman novelist". Born in 1752, she was the daughter of Dr. Burney, the well known music teacher and music historian. In 1786 she was appointed a Lady-in-Waiting to King George III's Queen Charlotte. According to her Diary the King had, at dinner on Wednesday November

5th 1788 at Windsor Castle, "broken forth into a positive delirium, which long had been menacing all who saw him most closely; and the Queen was so overpowered as to fall into violent hysterics ….. a page said I must come at once to the Queen….. My poor Royal Mistress! Never can I forget her countenance…..pale, ghastly pale she looked….I gave her some CAMPHOR JULEP, which had been ordered her by Sir George Baker". It seems to have allayed pain without increasing the pulse rate very much. An increased dose produced a sedative effect. So it was in effect a narcotic like ALCOHOL or LAUDENUM. It was used to fight a cholera epidemic in the United states of America in 1832. Patients were told to take it three times a day in water. It was often used as a composing draught at night. The existence of such a label is recorded by Penzer. A label with this title is in the Marshall Collection. It is made of silver but unmarked, rectangular in shape with rounded corners, decorated with a double reeded border and measuring 2.5cm by 1.3cm, slightly smaller then the set if six medicinal labels in the Collection all of which measure 2.6cm by 1.4cm. An oblong unmarked small silver-gilt or perhaps gilt metal label with this title has also been noted, probably designed for use in a travelling box.

9.79　CAMPHORATED SPIRITS OF WINE (Medicinal)

This title has been found in full on a tiny silver rectangular label with cut corners measuring 2.6cm by 1.4cm made by Samuel Hayne and Dudley Cater in London in 1843, a typical travelling box label

(fig. 9.47). An example appeared in the H.E. Rhodes' Collection. For one Victorian lady it had as a companion EAU DE COLOGNE. See above CAMPH DROPS, CAMPHtd. SPIRIT and CAMPHOR JULEP.

Fig 9.47

9.80　CAPILARE (Medicinal)

Capillaire is the maidenhair fern, capillaries herba, used for making a restorative syrup flavoured with ORANGE FLOWER WATER from the mid-Eighteenth century. Daniel Hockley

Fig 9.48

made in London in 1818, before he emigrated to South Africa, a double deep rounded rectangular label for CAPILARE, which was said to be used for sweetening harsh tasting wine. For this reason, perhaps, the Marshall Collection has a bin label for CAPILLAIRE, numbered 441, although it could refer to storage of the syrup.

9.81　CAPILLAIRE (Medicinal)

This is the title of a bin label, see above CAPILARE.

9.82　CAPSICUM (Medicinal)

A small unmarked late Victorian rectangular silver label with a simulated fine wrigglework border engraved for CAPSICUM with an 1890s type silver chain may have adorned a bottle containing this remedy. The fruits of most species of CAPSICUM contain capsaicin (methyl vanillyl nonenamide) which is said to have been used as a circulatory stimulant and analgesic.

Fig 9.49

9.83　Carnation (Perfume)

This title refers to the perfume and not to the baby-milk or milk based food for the infirm. The title is hand-written in script format as one of a set of four slot labels fitting into art nouveau or art deco pewter labels. The other titles are Sweet Pea, Lily of the Valley and White Lilac. The title also appeared on an oblong label. Curiously it also appears in the Master List. Slot labels for perfumes were also made by Britton Gould & Co. in silver for Messrs. J. Floris in 1905 of similar design and shape (see below EAU DE TOILETTE VERVEINE).

Fig 9.50

Fig 9.51

Fig 9.52

Fig 9.53

9.84 CASTOR OIL (Medicinal)

A triangular shield shaped enamel label in the Sanderson collection of the Worshipful Company of Vintners in Vintners Hall bears this title which may evoke painful childhood or childbearing memories. The oil is extracted from the seed of the castor oil plant Ricinus communis or the Palma Christi. It was known and used in ancient Egypt. From these early times it was used to deal with eye infections and irritations. It was then used, certainly in mid-Victorian times, for the relief of back pain by massage or by the application of a hot pad or pack or poultice soaked in the oil, which then penetrated the skin. It was also used to encourage the onset of labour in childbearing, to deal with constipation and for certain other abdominal complaints. Another example appeared in the H.E. Rhodes' Collection.

9.85 CELLADON (Perfume)

See EAU DE CELLADON below

9.86 CHARTREUSE (Medicinal)

White CHARTREUSE was made by the monks of a Carthusian Monastery near Grenoble to operate as a kind of BALM, which see above. It became a popular liqueur and was drunk in the boudoir, perhaps for its medicinal properties, along with COGNAC and ANNISETTE. The labels are highly decorative. Chartreuse in lower case lettering has a stove enamel backing and thus could have been made by Delvaux of Paris (figs. 9.51, 9.52 and 9.53).

9.87 CHERRY (Medicinal, Soft Drink)

This fruit imbued cordial (33), which gets better with age, was the subject of a silver label made by Peter, Ann and William Bateman in 1799 of the cupid holding a swag design (fig. 9.54). Another example, a pair with GINGER (34), was made by Crichton Brothers in London in 1953, bearing the Coronation mark, the Queen's head crowned and maker's mark LAC (also illustrated in fig. 9.54) being

the initials of Lionel Alfred Crichton, the founder of the firm. Another example was painted in gold on tortoise shell (35) of appeal to the ladies. This non-alcoholic beverage should be distinguished from CHERRY:BRANDY, CHERRY.BY, BOUNCE, GIN, MADEIRA, RATAFIA and WHISKEY, all in the Master List. A plated CHERRY (fig. 11.3) is in set with CURRANT, ORANGE and RAISIN. CHERRY was drunk by those having bronchial problems. It was also regarded as being good for the protection of teeth and gums. So it was retailed by Breiden Bach & Co., perfumers and distillers, of 127 New Bond Street, London. CHERRY could also have been the title for the liqueur Maraschino which was made from distilled cherry juice. The all America drink, CHERRY Bounce had sugar and vodka added.

Fig 9.54

9.88 CHIPRE (Perfume)

This was a popular perfume made by Guerlain from about 1909. It was also a Cyprus wine. If in silver the

Fig 9.55

size of the label determines. An enamel escutcheon shaped label for CHIPRE, however, measures 6.4cm by 4.7cm. It is in the Marshall Collection, numbered 418. An enamel label is illustrated (fig. 9.55).

9.89 Chloroform (Medicinal)

This title appears in blue script upon an oval porcelain label of fairly large size (fig. 9.56). It is a thin colourless liquid, a chloride of formyl, having a sweetish taste.

Fig 9.56

It gives off a vapour which, if inhaled, sends one off to sleep. Dr. John Collis Brown's version suppressed coughing. Other versions to soothe the nerves once included a mixture with cannabis and a product called chlorodyne.

9.90   CHYPRE (Perfume)

This was a popular perfume made by Guerlain from about 1909. It was also a Cyprus wine. This was an alternative spelling of CHIPRE which see above. The Bayeux pottery label displayed in the Bayeux Museum made by M.F. Gosse around 1860 with its title CHYPRE in black lettering was probably a wine label.

9.91   CINNAMON (Medicinal and Perfume)

Known to pharmacists as Cortex Cinnamoni, its botanical name is Cinnamomum Zeylanicum or Cinnamomum Verum. CINNAMON oil is a pure, natural essential oil. As a sauce label, the bottle so labelled contained a spice powder which worked as a kind of spicy sauce. As a perfume label, the bottle so labelled contained a strongly aromatic perfume. As a medicinal label, the bottle so labelled contained CINNAMON buds which look like cloves and were used as a substitute for cloves. The buds were unripe fruits harvested shortly after the plant had blossomed. For this reason CINNAMON has been found to be a pair with BORACIC. Cinnamon was well known to the Ancient Egyptians. A paper label on an Augsburg medical box of 1630 reads "Cinnamon Essence" in company with "Rose Water" and "Spirit of Sweet Balm". With regard to its use as a perfume, CINNAMON has been found in set with COLOGNE, EAU DE ROSE and ROSEWATER. It was also used as an important cordial in Holland around 1740 being the centrepiece of a 13-bottle travelling cordial set. An unmarked octagonal label is in the Anderson Collection. As a boudoir label it could well have related to a liquid used for anti-clotting, anti-microbial and blood sugar control purposes. Cinnamon quills were imported from Sri-Lanka and cinnamon leaves were sometimes used as a substitute for Indian bay leaves.

Fig 9.57

9.91a   CINQ A SEPT (Medicinal)

Fig 9.57a

This affectionate boudoir label, as indicated by the heart which stands for love, stood for five unto seven, which over time has come to mean the two hour semi-formal social gathering which sometimes took place in Quebec after work but before dinner. In Canada it seems wine, beer, spirits and cocktails are served with finger foods, perhaps not unlike an English wine and cheese party. But in France it stood for something quite different. It was a relaxant. According to a November 1966 edition of Time magazine "in the days of Maupassant, mustaches and mistresses, the affluent Frenchman could not do without his Cinq a Sept for an early evening liaison". So it seems that originally Cinq a Sept stood for time spent with a mistress and that this is the real significance of the heart device shown on the front of this well-worn label (see fig. 9.57a)

9.92   CITRATE MAGNESIA (Medicinal)

This formed one of a probable set of five plated labels perhaps of post World War I date shown to the Wine Label Circle by Mr. Clemson-Young. It was used for the relief of occasional constipation, usually producing a bowel movement within hours. This laxative was taken orally, but was often lemon flavoured to make it taste better. It was marketed at a later date under the name "Milk of Magnesia". See also MAGNESIUM CITRATE below.

9.93   CITROEN (Medicinal and Perfume)

See CITRON below. Barbadus Citroen was the title of one of the thirteen cordial bottles in a Dutch travelling case of around 1740. Citron is called citroen in Dutch.

9.94   CITRON (Medicinal and Perfume)

CITRON is the fruit of the tree Citrus Medica. It is like a lemon or lime but less acid. Citrus water

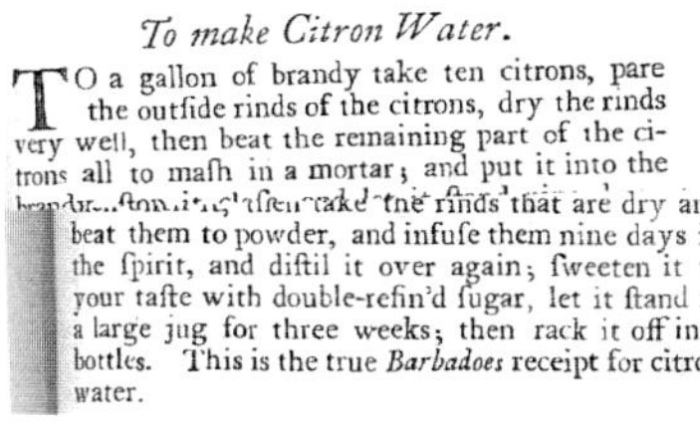

To make Citron Water.

TO a gallon of brandy take ten citrons, pare the outside rinds of the citrons, dry the rinds very well, then beat the remaining part of the citrons all to mash in a mortar; and put it into the brandy... then, after take the rinds that are dry and beat them to powder, and infuse them nine days in the spirit, and distil it over again; sweeten it to your taste with double-refin'd sugar, let it stand in a large jug for three weeks; then rack it off into bottles. This is the true Barbadoes receipt for citron water.

Fig 9.58

is a drink flavoured with CITRUS or LEMON peel. Citronella is an essential oil used for soap making and scenting hair-oil. POMERANS, a variety of citrus fruit, aurantium Amora-Gruppen, is used for skin care. There was also a perfume made from around 1800 known as Attar of Citron. So CITRON was a title on a splayed hoop silver neck ring label in the ATTAR OF ROSES set of 8 perfume labels. A recipe for making citron water (see fig. 9.58) is to take 10 citrons, pair the outside rinds, dry the rinds very well then beat the remaining part of the citrons all to mash in a mortar and then place them in a gallon of brandy. Infuse them 9 days in the spirit and distil it all over again. Sweeten it to your taste with double refined sugar, let it stand in a large jug for 3 weeks then rack it off into bottles. This is said to be the true barbadoes receipt for citron water. The labels would have been hung on the bottles for identification purposes. Citron water, a non-alcoholic cordial, was retailed according to his trade card by Rone (originally from Dublin) from near St. George's Church in Southwark around 1740.

9.95   CITRONELLE (Perfume)
See above, CITRON. Citronella oil is a pure, natural essential oil. CITRONELLE is the French name for a herb commonly known as balm gentle. CITRONELLE was included in the set of eight silver perfume labels, see above ACQUA D'ORO. It apparently had a lemony flavour as a perfume. Balm gentle was also used as a French cordial, for example in the making of CITRONELLE Ratafia, for which a recipe is given by Richard Dolby in his Dictionary.

9.95a   CITRONSAFT (Soft Drink)
A continental silver label with this title for LEMON JUICE was part of Lot 316 in Christie's sale of 5th November 2002. It was not illustrated. It was in set with HALLONSAFT (CHERRY CORDIAL) and KORSBARSSAFT (RASPBERRY AND REDCURRANT CORDIAL).

9.96   CLARY (Medicinal)
CLARY, sometimes called CLARY sage, is a biennial herbaceous plant known as salvia sclarea. This plant has a long history as a medicinal herb, grown especially for its essential oil. Descriptions of its medicinal use have been given by Theophrastus in the Fourth Century BC, by Dioscorides, by Pliny the Elder and more recently by Nicolas Culpepper in his "Complete Herbal" of 1653. The oil is used in aroma

Fig 9.59

therapy for relieving fear and anxiety and inability to sleep. In 1694 "The Ladies Dictionary: Being a General Entertainment for the Fair Sex" recommended a bath of sage mixed with claret, wormwood, chamomile and squinath. It also recommended an interesting slimming lotion : "Take an ounce and a half of oyl of foxes, oyl of lilies and capons grease … boil the brew in an earthenware pot, adding an ounce of oyl of elder". A Phipps and Robinson example of CLARY is illustrated, made in 1791 (fig. 9.59).

9.97   CLOVE WATER (Soft Drink and Toiletry)
This was a CORDIAL used in the boudoir. It was one of the thirteen titles on cordial bottles contained in a Dutch travelling case of around 1740. Cloves were analgesic, tending to remove pain.

9.98   CLOVES (Medicinal, Perfume and Toiletry)
Clove bud oil and clove oil are both pure, natural essential oils. Dr. Scott gives a recipe in 1826 for oil of CLOVES being used in a tooth-powder made up of orrice root, cuttle-fish bones and cream of tartar. Cloves were used to alleviate toothache. They were included in CORDIAL sets and in ladies' dressing cases. CLOVES (Syzygium aromaticum) are the dried flowers, known to pharmacists as Flores caryophylli, from the tree known botanically as Caryophyllus aramaticus. CLOVES was a title on a splayed hoop silver neck ring label in the ATTAR OF ROSES set of 8 perfume labels. The fact that it was used in quantity is shown by the larger size porcelain escutcheon label decorated with four red roses with foliage on a white background (fig. 9.60), and the large Sheffield plated labels, one measuring 5.0cm by 2.9cm in the Marshall

Fig 9.60

Fig 9.60a

Collection (numbered 40), where it is suggested that its storage jar was for the Cook's use below stairs, kept in the larder. One is illustrated in fig. 9.60a measuring 5.2cm by 3.0cm. Cloves were antiseptic, preventing decay.

9.99 COGNAC (Medicinal)

Rawlings and Summers around 1830 made a very small label with this title and William Summers later on in 1877 made a small plain crescent shaped label with this title. Neither label would have been for use in the dining room. They were probably made for the travelling box where labelled glass bottles contained restoratives. See BRANDY above for the full explanation. Interestingly COGNAC in set with NOYAU has been noted re-engraved over the titles CAYENNE and HARVEY on sauce labels made by Reily and Storer. COGNAC is also found in set with

ANNISETTE and CHARTREUSE having medicinal properties.

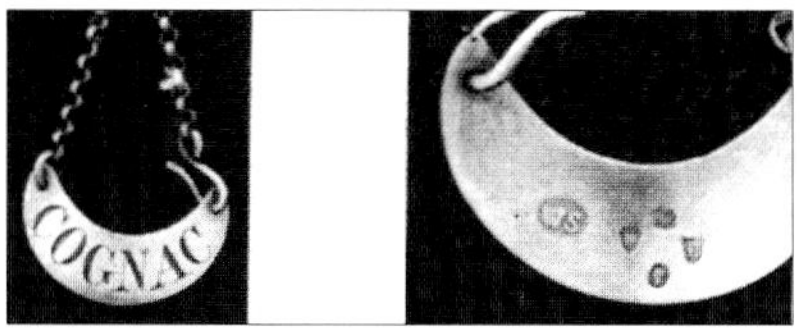

Fig 9.61

9.100 Cogniac (Medicinal)

This title in script appears upon a shield shaped label made in France in set with Limonade and Ratafiat de fleur d'Orange, on which see below and for COGNAC see above. The use was presumably for restorative purposes.

Fig 9.62

9.101 COLARES (Perfume)

An Irish label for COLARES was made in Cork by Toleken in a set with PARFAITE AMOUR and three others around 1795. Susanna Barker also made a silver COLARES label around 1780 (maker's mark only) which measures 4.5cm by 0.9cm and is numbered 41 in the Marshall Collection.

9.102 COLLARES (Perfume)

See COLARES above.

9.103 COLLICK WATER (Medicinal)

This is an organic gripe water relieving colic discomforts. It has been used by mothers for generations to provide relief from stomach cramps, hiccups, flatulence and teething. Other traditional

natural remedies such as fennel and ginger are contained in gripe water made for use by infants. Ginger helps to relieve nausea and other digestive problems, fennel is effective for relaxing a baby's intestinal tract and sodium bicarbonate is used to decrease stomach acidity.

9.104 COLOGNE (Perfume)

See below, EAU DE COLOGNE. It was reported in the Journal that "a German COLOGNE label with reclining cupids bought at the historic Covent Garden perfumers, Penhaligan's, was displayed at the Circle's AGM in 1978".

9.105 COLORLESS HAIR TONIC (Toiletry)

An essential label for the boudoir, this title was engraved upon a plain reeded octagonal silver label made in the USA. It is marked "STERLING 10" and dates from the 1930s (fig. 9.63). Hair tonics with a variety of ingredients have been used for years. Some are based on Tokara sea-water, taken from the islands lying between the Pacific Ocean and the East China Sea, which is considered to be some of the purest water and richest in minerals deposited by the Japan current. For example, Noeviv Tonic, which is colourless, is said to be able to help the essential moisture to the hair and scalp. Some are based on herbs thought to condition and revitalise the scalp, being extracts from humulus lapulus, the root of coptis japonica, swertia japonica, the fruit of capsicum frutescens or the bark of betula alba.

Fig 9.63

9.105a Cooking Brandy (Medicinal)

See above C. PORT for explanation. This was a cardboard label in set with Damson Cordial and Ginger Wine. See also CORDIAL below.

9.105b Cooking Sherry (Medicinal)

See above C. PORT for explanation.

9.106 COPYING (Writing Ink)

This silver label is unmarked but it and its chain have been tested and found to be of sterling silver. It dates to around 1830, at a time when many small silver labels were not marked, and is said to be part of a set of Writing Ink labels along with BLACK and RED. Sir Walter Scott used GREEN for amending proofs.

Fig 9.64

Fig 9.65

BLUE, VIOLET and YELLOW inks were also made. COPYING Ink was a special ink used in conjunction with a copying machine where copies of a writing needed to be made by impression.

The label is fitted with a slip chain for easy cleaning. The title is bold and easily read for speedy use with a campaign chest's copying machine when urgent orders needed to be copied for commanders in the field of battle. COPYING ink according to Norman Henley's "Twentieth Century Formulas, Recipes and Processes" is prepared by adding a little sugar to ordinary BLACK ink, which for this purpose "should be rich in colour". Writing executed with this ink may be copied "by passing it through a copying press in contact with thin, unsized paper, slightly damped". He gives recipes for three inks, to be tried out using an ordinary steel pen to check that the mixture flows freely: firstly, BLACK ink is usually made from nutgalls and a solution of salt and iron; secondly, RED ink was once upon a time made from carmine or cochineal but this has been superseded by potassium eosin; and thirdly, YELLOW ink is usually made from an aniline dye. Henley also gives detailed instructions for the preparation of COPYING ink for use without a press.

9.107 CORDIAL (Medicinal and Soft Drinks)
Some CORDIALS were used as pick-me-ups. A recipe (see fig.9.66) for a fine Cordial-water shows that it was brandy based. MELOMEL for example (see below) was based on cordial with added blackberries mixed with honey wine and was popular with the ladies. So also was a red cordial made from raspberries mixed with water and possibly non-alcoholic. CORDIAL is mentioned by Penzer. Thomas Watson in 1663 even wrote a theological treatise entitled "A Divine Cordial". Having noted that Ambrose had said that the whole of scripture should be the feast of the soul, Watson explains that Romans Chapter 8 should be a dish at that feast and

"like a cordial with its sweet variety" should refresh and animate the hearts of God's people. The title appears on a silver broad rectangular label made by William Troby in London in 1828 (Fig. 9.65). Other kinds of cordials known include: burgundy, gin (see below), ginger (see below), chambertin and damson (with ginger wine and cooking brandy being cardboard labels on boxed decanters). An enamel CORDIAL is illustrated by figure 1025 in "Wine Labels". In Scotland Robert Gray produced a CORDIAL label measuring 3.5cm by 1.9cm and made in Glasgow but marked in Edinburgh circa 1805 which is shown hanging on a flask in fig. 14.1. There is an example in Sheffield plate in the Marshall Collection numbered 43 and measuring 5.7cm by 3.1cm.

A fine Cordial-water.

BEAT two pounds of double-refin'd sugar very well, and put to it a gallon of the best brandy, stirring it a good while all one way; then put coriander of alkermes one dram, oil of cloves one dram, spirit of saffron one ounce, then stir it one way for quarter of an hour, then add three sheets of leaf gold and bottle it up, it will keep as long as you please.

Fig 9.66

9.108 CORDIAL GIN (Medicinal)
A silver label with this title was made in Birmingham in 1830 by Joseph Willmore measuring 4 x 2.2cm. It is numbered 44 in the Marshall Collection. It was used as a restorative. See above, CORDIAL.

9.109 CORDIALS (Medicinal and Soft Drink)
See CORDIAL above. CORDIALS of various kinds were carried in picnic and other kinds of boxes. Cordials became popular in the Eighteenth Century as a great aid to digestion. A silver label is illustrated (fig. 9.65 front and back), along with a recipe (fig. 9.66) for a fine Cordial-water.

9.110 COWSLIP (Medicinal and Soft Drink)
COWSLIP was prescribed for insomniacs and taken as a night cap. Around 1805 Elizabeth Morley made a plain cut-cornered rectangular silver label attractively pierced for COWSLIP. The popularity of COWSLIP is shown by the broken bin label with this title numbered 446 in the Marshall Collection. It has the Wedgwood mark stamped on its back. Home brews were stored. An old Sheffield plate neck-ring for COWSLIP shows signs of heavy wear (36).

Another Sheffield plate example is in the Marshall Collection numbered 45, measuring 4.4cm by 2.6cm which would be appropriate for a boudoir label. Lea & Co. of Birmingham made a tiny plain octagonal COWSLIP label in 1821 which may have been used in the boudoir (fig. 9.67).

9.110a COWSLIP WINE (Medicinal)
COWSLIP WINE is still a favourite home brew – see for example Eric Durkin's COWSLIP WINE designed by Harold Baker in the 1970s and retailed in 1973 by Payne & Son from their shop in the High at Oxford. It was an appropriate boudoir drink, taken medicinally.

9.111 CRAB APPLE (Soft Drink)
Crab is the name given to the wild apple, especially connoting its sour, harsh quality. A restorative fruit juice was made from crab apples. CRAB APPLE is part of a five piece boudoir set of labels made in 1956 by Wakely and Wheeler and retailed in London by Garrard & Co. The labels are in silver-gilt and are cast with a decoration of various fruits, flowers and bees (fig. 9.68).

9.112 CRÈME DE CACAO (Soft Drink)
See above CACAO.

9.113 CRÈME de CAFÉ (Perfume)
This title suggests a stimulant perhaps taken after a heavy meal, but in fact it refers to a perfume as the label is one of the ACQUA D'ORO set of eight silver labels. It was also used to give flavour to a commercially made liqueur.

9.114 CRÈME DE FLEURS D'ORANGE (Toiletry)
This was a toilet preparation scented with the perfume prepared from orange flowers – see further below, FLEUR D'ORANGE.

9.115 CRÈME DE NOYAU (Medicinal and Perfume)
See below, CRÈME DE NOYEAU.

9.116 CRÈME DE NOYEAU (Medicinal and Perfume)
This title is also related to perfumery. CRÈME DE NOYEAU ROUGE is one of the ACQUA D'ORO set of eight silver labels. Labels for Crème de Noyeau were made in London in 1792 and also in London

Fig 9.67

Fig 9.68

Fig 9.69

by Joseph Price in 1832 as a pair with Crème de Thé (fig. 9.69), and by John Heath and John Middleton of Hukin & Heath in 1908 (Marshall Collection, number 47).

9.117 Crème De Noyeau (Medicinal and Perfume)
It is paired with a silver label made by Joseph Price in London in 1832 also engraved in script for Crème De Thé, both in the Anderson collection. It was a perfumed soft drink and possibly a pick-me-up as well. See Crème De Thé below.

9.118 CRÈME de NOYEAU ROUGE (Perfume)
This title refers to a perfume being one of the ACQUA D'ORO set of eight silver labels.

9.119 CRÈME DE PORTUGAL (Perfume)
See below, CRÈME DE. PORTUGALE.

9.120 CRÈME DE. PORTUGALE (Perfume)
This eighteenth century French enamel, made circa 1770, and shown in fig. 9.70, is in set with HUILE D'ANIS ROUGE and EAU. D'OR. Of escutcheon shape with a creamy white background, the label has a pink rose in a floral display above the title but also displays a bunch of bluish grapes below the title. This set of pre-Samson French enamels has labels with blue eyelets. It comprises some of the smallest size of labels noted, which is appropriate for perfumes.

Fig 9.70

9.121 CRÈME DE PORTUGAU (Perfume)
See above CRÈME DE. PORTUGALE

Fig 9.72

Fig 9.73

Fig 9.74

Fig 9.71

9.122   CRÈME DE ROSE (Perfume)
This perfume is a current product of Yves Rocher.
It is used to flavour a commercially made liqueur
and thus has given the liqueur its name. See further
above ATTAR OF ROSES and C.DE ROSE.

9.123   CREMEDEROSE (Perfume)
See above, CRÈME DE ROSE.

9.124   Crème De Thé (Medicinal and Soft Drink)
See Crème De Noyeau above. This was its pair
with a script title made by Joseph Price in 1832 (fig.
9.71). Tea was very much a boudoir drink. This
was a travelling box label which from one corner
kept company with a perfumed refresher in the
other corner of the box. Its use as a pick-me-up is
evidenced by an article published in The Times on
Prohibition which describes Crème De Thé as an
Eighteenth Century rare spirit.

**9.125   CRÈME DE VANILLE (Perfume and
          Toiletry)**
This is a perfumed body moisturising cream.
William Summers in 1859 produced an oblong
gadrooned silver label with this title. A Samson
enamel label (fig. 9.72) entitled CRÈME DE
VANILLE, of similar design and appearance to
WHITE LILAC, WHITE ROSE and VIOLET, is of a
size suitable for the lower scent standard toilet water
bottle rather than the higher standard scent and
smaller perfume bottle, being approximately 5.7cm
by 3.9cm. CRÈME DE VANILLE is also the name
given to a commercially produced liqueur.

9.126   CRÈME DE VIOLETTES (Perfume)
This is a strong perfume. For details see VIOLET
below.

9.127   CURRANT (Soft Drink)
A number of soft drink labels were said to have been

represented in the Alexander
Cuthbert Collection
(37). The list probably
included CURRANT,
RED CURRANT, RAISIN,
GOOSEBERRY and ELDER.
However the larger labels
were used probably to identify
wines rather than soft drinks.
CURRANT and RAISIN
were popular wines. The large
Battersea enamel (fig. 9.77)
and the large Staffordshire
enamel shown in fig. 9.77c
were probably for wines.
However the two labels for
CURRANT in the Marshall
Collection, numbered 48
and 49, measure 4.1cm
by 2.0cm and 4.5cm by
2.4cm respectively, one
being made by Samuel
Godbehere, Edward
Wigan and James Bult
in 1804 and the other by
Solomon Hougham in
1812. Another smallish
but early label (c1765)
by WM (attributed to
William Moody) has
been illustrated in the
Journal (fig. 9.78). An
example illustrated
(fig. 9.75) by Thomas
and James Phipps,
made in London in
1816, is in pair with
GRAPE and probably in

Fig 9.75

Fig 9.76

Fig 9.76a

Fig 9.77

Fig 9.77a   Fig 9.77b

set with PORT and SHERRY since all four labels
are marked with the owner's initials "G.S.B." but of
course the labels could have been used in pairs. The
plated CURRANT (fig.9.76 and see fig. 11.3)) is
of small sauce label size. A silver version is a little

Fig 9.77c

Fig 9.78

Fig 9.77d

larger. Also shown in are two octagonal labels, the upper by Thomas and James Phipps in 1817 and the lower by Lea & Co probably in 1821 (figs. 9.77a and 9.77b). Four other examples are illustrated, a double-reeded curved octagonal by Joseph Willmore of Birmingham dated 1816 (fig. 9.73), an elongated beaded kidney by Cocks and Bettridge of Birmingham in 1814 (fig. 9.74), a double-reeded rounded-end rectangular by William Constable of Edinburgh in 1807 assayed by Ziegler (fig. 9.76a) and a double-reeded long narrow octagonal by Joseph Willmore in Birmingham, fancily marked in the shape of a cross, made in 1811 (fig. 9.77d).

9.128 CURRANTS (Soft Drink)
This title appears upon a broad rectangular silver label made by Elizabeth Morley in London in 1811. See above, CURRANT.

9.129 DACTILLUS (Perfume)
A Samson enamel label, a pair with WHITE ROSE, very small being approximately 3.0cm by 2.6cm, has this tantalising title. A pterodactylus was, for example, a dinosaur. However, according to Hartlief's "Biologiezentrum", DACTILLUS was a herb, which presumably would have been distilled to make perfume. It would be a suitable adornment for a perfume bottle in a travelling case but being an enamel it was more likely to have been hung round a cut glass bottle on the dressing table. Unusually for a Samson label it has no dot over the capital I (fig.9.79) nor has its pair and nor have three other Samson WHITE ROSE labels all of similar but not identical design.

Fig 9.79

Fig 9.80

9.130 DACTILUS (Perfume)
See above, DACTILLUS.

9.131 DAMSON (Medicinal and Soft Drink)
A cardboard label for Damson Cordial forms part of a 1780's decanter travelling box, perhaps used in connection with picnics, in set with Ginger Wine and Cooking Brandy. An example in old Sheffield plate is a pair with GOOSEBERRY (which see below) and an enamel example is illustrated (fig. 9.80).

9.131a Damson Cordial (Medicinal)
In set with Cooking Brandy and Ginger Wine, which see. These were detachable cardboard labels. See also CORDIAL above.

9.132 DANDELION (Soft Drink)
For a long time the common dandelion (Taraxacum officinale) has been cultivated for use as a herbal drink. An example of a DANDELION label in silver of elongated escutcheon form is double sided with RASPBERRY. It was made by Charles Davis in London in 1960.

9.132a DANDYLION WINE (Medicinal)
DANDELION was also a popular home-made country wine: see, for example, Eric Dorkin's 1970s DANDYLION WINE, made to the design of Harold Baker and retailed by Payne & Son of Oxford in 1973. Various recipes exist for wine making using only the petals of the flowers.

9.133 Delicieux (Perfume)
This tiny unmarked silver perfume label with its title in script was shown on Antiques Roadshow on 23rd October 2005 and would seem to date back to the 1750s. The Household Cyclopedia of Perfumery gives a formula for making Bouquet des Delices (which was also known as Bouquet de Caroline). Pierre-Francois-Pascal Guerlain retailed Delice de Prince in Paris from 1863.

9.134 DENTIFICE WATER (Toiletry)
See below, DENTIFRICE

9.135 DENTIFRICE (Toiletry)
This toilet preparation, known also as DENTIFRICE WATER, EAU DENTIFRICE, EAU DENTIFICE and TOOTH MIXTURE, could thus be in liquid or powder form. It was used for cleansing one's teeth by rubbing or brushing and has certainly been in use

since Tudor times. Cleaning teeth is an essential part of preventing tooth decay and thus DENTIFRICE was an essential component of any toilet box, presumably instantly recognisable through constant use and not requiring a label except in the bathroom to avoid confusion with other preparations kept there.

9.136   DENTIFRICE WATER (Toiletry)
See above, DENTIFRICE. The title appears on a silver label made in 1834 by Bent and Tagg in Birmingham. It is numbered 388 in the Marshall collection and is octagonal in shape with triple reeding, measuring 4.7cm by 2.5cm.

9.137   DENTRIFICE (Toiletry)
See above, DENTIFRICE

9.138   DENTRIFICE WATER (Toiletry)
See above, DENTIFRICE

9.139   DISTILLED WATER (Soft Drink and Toiletry)
It used to be said by some that the process of distillation takes out of the water all the beneficial minerals. What it does is to remove a range of contaminants so that it comes closest to the definition of pure drinking water. It will kill and remove bacteria and viruses as well as heavy metals and radionuclides. When drunk, DISTILLED WATER picks up rejected and discarded minerals and with help from blood and lymph take them to the lungs and kidneys for elimination from the body (38). The original distillers did not incorporate in their process any pre or post carbon filtration. This tended to give DISTILLED WATER a flat, steamy or "off" taste. Its labelled container would be taken out of the travelling box and placed on a bedside or dressing room table for overnight use. So Rawlings and Summers made a substantial silver rectangular with rounded ends label for this purpose in 1838 with an engraved title and distinctive treble reeded *Fig 9.81a*   border (fig. 9.81a). It has the short lived

head of the poorly engraved young Queen Victoria (the engraver was going blind) as the duty paid mark which was replaced after Wyon won the successful competition with a new punch. See fig. 9.81b.

9.140   D'ORGEAT (Medicinal)
See Orgeat below.

9.141   Drinking Water (Soft Drink)
This title was engraved in script on a narrow rectangular label with rounded ends of silver appearance but probably plated. It was unmarked and first noted by Major Gray in 1958 (39). It was not curved for an elegant decanter in the dining room. It was therefore probably hung on a square bottle which would happily pack into a large travelling box or look well in a boudoir as indeed would EAU, MALVERN WATER and DISTILLED WATER.

9.142. EAU   (Soft Drink)
An example in ivory of hoop or neck-ring design is in the Cropper Collection at the Victoria and Albert Museum. See also above, Drinking Water.

9.143   EAU ASTRINGENTE (Medicinal and Toiletry)
This lotion was used by ladies to close up unwelcome open pores on their faces, thus maintaining a good complexion. See above ASTRINGENT.

9.144   EAU BORIGUÉE (Medicinal and Toiletry)
This is a boracic lotion for the eyes. The title appears upon a French enamel label (fig. 9.82) surmounted by a ribbon in mauve colour.

9.145   EAU BORIGUEES (Medicinal and Toiletry)
See above EAU BORIGUEE.

9.146   EAU BORIGUES (Medicinal and Toiletry)
See above EAU BORIGUEE.

Fig 9.82

Fig 9.81b

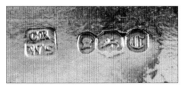

9.147 EAU BORIQUEE (Medicinal and Toiletry)

See above EAU BORIGUEE and BORACIC WATER. The title EAU BORIQUEE appears on a French enamel label with floral decoration in set with EAU DE TOILETTE and ELIXIR DENTIFRICE, of the usual escutcheon shape, measuring 5.6cm by 3.8cm or thereabouts. This set is in the Marshall Collection, numbered 420 to 422.

9.148 EAU D'ALIBURGH (Medicinal)

This was a remedy for scabies which is a general term for skin diseases characterised by the appearance of scabs or scales. Such diseases are contagious and itch like mad causing one to want to scratch. They are caused by the parasite sarcoptes scabei. Presumably Eau D'Aliburgh, sometimes called D'alibour, soothed the skin as it was applied by a compress. An unmarked silver label with this title forms part of a set of six medicinal labels in the Marshall Collection numbered 344 to 349.

9.149 EAU D'ARQUEBUSE (Medicinal)

Otherwise known as ARQUEBUSADE, for which see above. Eau d'Arquebusade was a product of Pierre-Francois-Pascal Guerlain retailed in the Hotel Meurice, Rue de Rivoli, Paris.

9.149a EAU·DE·CAFFÉ (Perfume)

Fig 9.82a

Believed to be known also by the name Crème de Café, a label with this title being in a set of eight perfume labels, this perfume was used to flavour a liqueur (fig. 9.82a).

9.150 EAU DE CELADON (Perfume)

Lady Primrose's Celadon is a famous perfume or Eau de Parfum still available for purchase today. The title appears upon a Staffordshire enamel escutcheon datable to the 1780's. It appears as a pair with SCUBAC sold at Christie's South Kensington on 12th June 1980 (40). The title also appears in Major Gray's list of April 1958 (39). CELADON is a pale green colour. HM The Queen wore a pale celadon green outfit at the opening of Royal Ascot on 14th June 2011. Chambers Dictionary suggests the colour was named after a character in D'Urbe's "Astree".

9.151 EAU DE CELLADON (Perfume)

See EAU DE CELADON above.

9.152 EAU DE COL (Perfume)

See EAU DE COLOGNE below. See also EAU DE COLODON below. The silver EAU DE COL label is numbered 611 in the Marshall Collection and dated 1829. It is a pair with TOOTHMIX and from its size it would appear that this pair graced bottles in a travelling box, being only 2.2cm by 1.1cm. The labels are said to have been made by Thomas Streetin.

9.153 EAU DE COLODON (Perfume)

This is a French perfume made at La Harth which is near Mulhouse from the flowers of the Hieracium Colodon. It is in set with Huile de Vinus being taken from a casket or case containing a set of six 18th century glass perfume bottles all sold at Sothebys on 15th February 1979. Each of the six bottles had an unmarked oval shaped silver label.

9.154 EAU DE COLOGNE (Perfume)

This very popular toilet water in to-day's times usually has an essential oil concentration of around five percent. However, it was originally a perfume of light, fresh fragrance, mixed with citrus oils, known as Aqua Admirabilis, created around 1709 by an Italian barber from the Val Vigezzo called Gian Paolo Feminis. He lived in Cologne (Koln) in Germany at this time. In 1732 his nephew Giovanni Maria Farina took over the business. In due time the formula ended up with the perfume house of Gallet in Paris. Examples of labels bearing this title include (i) a pink French Samson enamel made around 1890 (fig. 9.86d), (ii) a silver neck-ring label by Thomas Diller, sometimes attributed to Thomas Dicks (Grimwade 2731), dated 1837 (figs. 9.84a and 9.84b), (iii) a Victorian silver label by George Lowe of Chester in 1838 marked on the face (fig. 9.85), (iv) a pink floral Samson enamel (compare figure 1015 in "Wine Labels") shown in fig.9.86e, (v) a heart-shaped porcelain decorated with pink roses (fig. 9.86f), (vi) a small double-reeded rectangular rounded ended silver-gilt label made by John Reily with pierced lettering, (vii) a shell and scroll silver-gilt label by Charles Rawlings circa 1820 (fig.9.86c), (viii) a Crown Staffordshire English porcelain with the title in gold lettering, pattern number A 6297/1 (fig. 9.83), (ix) an oval shaped enamel with scrolls and red and yellow roses (fig. 9.86), (x) a silver-gilt leaf design label of 1834

Fig 9.83

Fig 9.84

Fig 9.84a

Fig 9.84b

Fig 9.85

Fig 9.86

Fig 9.86a

Fig 9.86b

Fig 9.86c

Fig 9.86d

Fig 9.86e

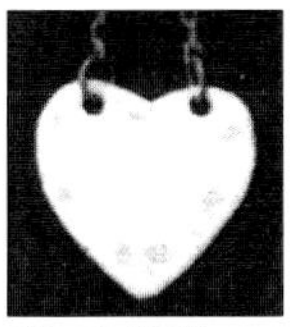

Fig 9.86f

Fig 9.87

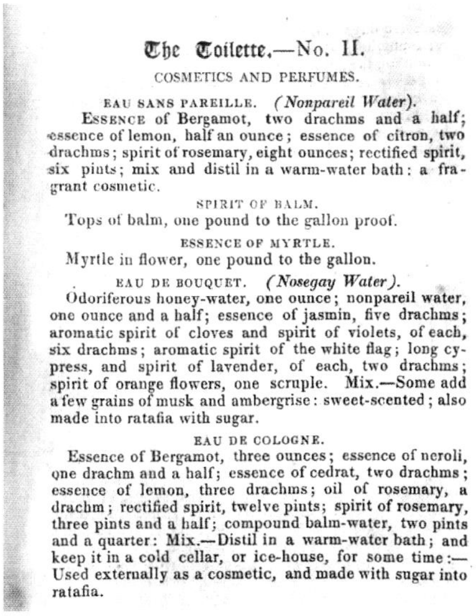

The Toilette.—No. II.
COSMETICS AND PERFUMES.

EAU SANS PAREILLE. *(Nonpareil Water)*.
Essence of Bergamot, two drachms and a half; essence of lemon, half an ounce; essence of citron, two drachms; spirit of rosemary, eight ounces; rectified spirit, six pints; mix and distil in a warm-water bath: a fragrant cosmetic.

SPIRIT OF BALM.
Tops of balm, one pound to the gallon proof.

ESSENCE OF MYRTLE.
Myrtle in flower, one pound to the gallon.

EAU DE BOUQUET. *(Nosegay Water)*.
Odoriferous honey-water, one ounce; nonpareil water, one ounce and a half; essence of jasmin, five drachms; aromatic spirit of cloves and spirit of violets, of each, six drachms; aromatic spirit of the white flag; long cypress, and spirit of lavender, of each, two drachms; spirit of orange flowers, one scruple. Mix.—Some add a few grains of musk and ambergrise: sweet-scented; also made into ratafia with sugar.

EAU DE COLOGNE.
Essence of Bergamot, three ounces; essence of neroli, one drachm and a half; essence of cedrat, two drachms; essence of lemon, three drachms; oil of rosemary, a drachm; rectified spirit, twelve pints; spirit of rosemary, three pints and a half; compound balm-water, two pints and a quarter: Mix.—Distil in a warm-water bath; and keep it in a cold cellar, or ice-house, for some time :— Used externally as a cosmetic, and made with sugar into ratafia.

made by Rawlings and Summers in pair with EAU DE PORTUGAL (fig.9.86b), (xi) an initialled neck-ring (fig. 9.86a), (xii) another Crown Staffordshire "England" label pattern number A6297 but not having the title in gold lettering (fig. 9.84) and (xiii) an unmarked porcelain label (fig. 9.83). A recipe for making EAU DE COLOGNE was given by Dr. Scott in 1826 (fig. 9.87). EAU DE COLOGNE forms part of a three bottle perfume set, made by John Reily and numbered 334 to 336 in the Marshall Collection, along with ESSENCE OF ROSE and LAVENDER. EAU DE COLOGNE is a refreshing blend of rosemary, oil of neroli, bergamot, lavender and lemon. It was widely used during the seven years war, as a cure-all for a variety of illnesses or discomforts, often called at that time Admirable Water or Wonder Water. Home-made Cologne using published recipes was in vogue both in England and in America from the 1790's. It was also made popular by Napoleon. Elizabeth Raffald's "The Experienced English Housekeeper", published in 1799, included "near Nine Hundred Original Receipts, most of which never appeared in print". The Household Cyclopedia of Perfumery gives recipes for "J. Maria Farina Cologne", "Fine Cologne Water", "Ordinary Cologne", "Cheap Cologne" and "Cologne Water, from Redwood Gray's Supplement". Pierre-Francois-Pascal Guerlain retailed in Paris "Eau de Cologne Imperiale" from 1853 and "Eau de Cologne Russe" from 1863. EAU DE COLOGNE has also appeared in set with VERBENA, AROMATIC VINEGAR and MILLE FLEURS by Margaret Binley between 1764 and 1778. Lynas Perfumiers produced an EAU DE COLOGNE known as "Lady Luxury".

9.155 EAU.DE.COLOGNE (Perfume)
See above EAU DE COLOGNE

9.156 Eau De Cologne (Perfume)
See above EAU DE COLOGNE

9.157 Eau de Cologne ROSSE (Perfume)
Perhaps a rather special kind of EAU DE COLOGNE, for which see above.

9.158 EAU DE LAVANDE (Perfume)
See below, LAVENDER WATER

9.159   EAU DE LUBIN (Perfume)

This perfume was invented it seems around 1797 in France and became very popular at the end of the Nineteenth Century. Eugene Grasset produced a poster showing its use in the boudoir where it was described as the Queen of Toilet Waters (fig. 9.88b). The title appears on a handsome French enamel from the Samson factory surmounted by a pink bow (fig. 9.88a). The attribution is based on its size and shape (approximately 41 x 39.3mm) being similar to a Samson blank in the Victoria and Albert Museum. The French

Fig 9.88a

Fig 9.88b

provenance is supported by the language of the label and the dot inserted over the capital letter "I", and by the fact that the perfume house known as Lubin, founded in 1797, is based in Paris.

9.160   EAU de MIEL (Medicinal, Perfume, Soft Drink and Toiletry)

EAU de MIEL had a variety of uses. As a perfume it was produced by Guerlain in Paris from 1828. It was also sold as a toilet water which was considered to be a good hair-dressing. In a campaign chest made at the time of the Peninsular War it was paired with Arquebuzade, having been used as an anti-paralytic medicine. It was also used as a non-alcoholic cordial being in fact merely flavoured water

9.161   Eau de Nil (Toiletry)

This tiny gilt bronze label measures only 3.0cm by 1.5cm. Water taken from the River Nile (presumably the Upper Reaches) was thought to have been of benefit to the skin.

Fig 9.89

9.162   EAU DE NOYAU (Medicinal)

This title appears upon a crescent shaped French enamel. See EAU DE NOYEAU and NOYEAU below.

9.163   EAU DE NOYAUX (Medicinal)

This title appears upon a label made of mother-of-pearl. See EAU DE NOYEAU and NOYEAU below.

9.164   EAU DE NOYEAU (Medicinal)

See NOYAU, NOYEAU and NOYEO below. This title appears on a French enamel crescent style label and is mentioned by Penzer (41). It was a cordial and a pick-me-up made from peach and apricot kernels with the addition of prunes and celery. Sometimes BRANDY was added. It would have been carried in the travelling box. See NOYEAU below.

9.165   EAU DE NUIT (Soft Drink and Toiletry)

This label, rather on the large side, is illustrated in "Wine Labels" at figure 1194. It may have adorned a dressing table carafe (Fig. 9.90). Being made in electroplate it was

Fig 9.90

not for show in the dining room. A Samson enamel exists with the same title. Rainwater was not only drunk but also used for looking after the skin and complexion: see EAU DE PLUIE. Night-water is defined in the Oxford English Dictionary as water which is collected or stored during the night.

9.166   Eau de Nuit (Soft Drink and Toiletry)

This version (figs. 9.91 and 9.91a) appears on

Fig 9.91

Coalport bone china, "Made in England", established 1750 being the registered trade mark, with the crown in blue, contained in its original 1930's box (fig. 9.91b) inscribed "A gift for all

Fig 9.91a

Fig 9.91b

seasons" showing a lady in a yellow dress receiving a gift from her small son, pattern number 24.

9.167 EAU D'OR (Medicinal, Perfume and Soft Drink)

This is thought to be an abbreviation of EAU D'ORANGE, which see below, or to relate to GOLD WATER or GULDEWATER, about which also see below. The title appears engraved on possibly a re-used two putti die-stamped label by Joseph Willmore of Birmingham in 1814 (see fig. 9.92), with grapevine and mask of Bacchus decoration and a length of 5.9cm. (42) which seems somewhat inappropriate for such a delightful floral fragrance for women, although a German version of reclining cupids is known chosen for a COLOGNE label and so therefore be of feminine appeal. It also appears on an escutcheon shaped enamel label. Furthermore an enamel label with this title is in set with HUILE D'ANIS ROUGE and CRÈME DE. PORTUGALE, being the smallest known pre-Samson French escutcheons. They all have distinctive blue eyelets. See further below Fr d'Orange.

9.168 EAU. D'OR (Medicinal, Perfume and Soft Drink)

This title appears on an enamel version (fig. 9.93). See above, EAU D'OR.

9.169 EAU d'ORANGE (Perfume and Soft Drink)

See below Fr. d'ORANGE.

9.170 EAU D'ORANGE (Perfume and Soft Drink)

The title appears on an enamel label which has been illustrated in the Journal (43), in company with ANISETTE and other labels. See below Fr. d'ORANGE.

9.171. EAU DE PLUIE (Soft Drink and Toiletry)

Rainwater has always been regarded as being good for the complexion. It was certainly an inexpensive toiletry.

9.172. EAU.DE.PORTUGAL (Perfume and Toiletry)

As well as being a perfume compounded from oil of lemon, bergamot, sweet orange, otto of roses and alcohol, it was said to be a gentleman's hair tonic or lotion, good for restraining the onset of baldness. One variety of hair lotion was of French manufacture by Pinaud of Paris, a long established business with over two hundred years of experience. On a paper label on a bottle it is described as a lotion which stimulates surface circulation of the scalp. It removes loose dandruff, it keeps hair healthy, well-groomed and lustrous. The small oval silver label with this title and a thread border (fig. 9.94) is by William Stephen Ferguson of Elgin in Scotland of around 1830 in date. It is thought to have adorned a glass bottle in a perfume box. Ferguson moved to Elgin in 1828. Another variety was produced by Guerlain in Paris from 1828 under the title Eau de Portugal de Montpellier. See also EAU DE PORTUGAL below.

Fig 9.93

Fig 9.92

Fig 9.94

Fig 9.95

9.173 EAU DE PORTUGAL (Perfume and Toiletry)

This pierced title appears on a single vine leaf label on its side, a pair with EAU DE COLOGNE made by Charles Rawlings and William Summers in 1831. (Phillips Bond Street December 1985). This perfume is compounded from oil of lemon, bergamot, sweet orange, otto of roses and alcohol. It was very much in vogue in the 1830s. It was also manufactured by Pinaud of Paris and the name has appeared as a paper label on a dressing-table bottle containing a lotion said to stimulate the surface circulation. It is still in vogue and in use at the Cavalry and Guards Club in London where in the Gentlemen's Lavatory there is a bottle labelled "Eau de Portugal. Cologne. Made to the original formula by Mr. Philip, Gentleman's Hairdresser to the Club." Furthermore EAU DE PORTUGAL with or without oil is retailed to-day as a hair lotion by the firm of D.R. Harris in St. James's Street. See fig. 9.95 and EAU.DE.PORTUGAL above.

9.174 EAU DE ROSE (Perfume)

See below ROSE. This title appears upon (i) a Crown Staffordshire England porcelain with olive green markings and pattern number A 6297 in red; (ii) a middle size Dreyfous enamel; (iii) a Samson enamel which has been illustrated in the Journal (44) and (iv) upon various other labels including a fully marked silver-gilt London label of 1830 in the form of the rare Rococco cartouche with cherub's head design invented by Digby Scott and Benjamin Smith and copied (not necessarily in silver-gilt) by Charles Rawlings. The maker's mark is unfortunately obliterated. The most likely candidate is Rawlings and Summers. It is numbered 589 in the Marshall Collection and measures 3.0cm by 2.6cm.

Fig 9.96

Fig 9.97a

Fig 9.97b

Fig 9.97c

9.175 EAU DE TOILETTE (Perfume and Toiletry)

This is a general title covering a range of possibilities, clearly used in a toilet box or on the dressing table depending upon size. The name appears in Penzer's Table III , on a French enamel label with floral decoration of escutcheon form, in company with EAU BORIQUEE and ELIXIR DENTIFRICE (Marshall Collection numbered 421), on a Dreyfous French enamel label (45), on an oval enamel with red and yellow scrolls (46), on a small pink enamel and on a Samson label which exactly matches in size a Samson blank of unusual shape with a dot over the "i" of TOILETTE. Toilet water was less concentrated than eau de parfum. Tiffany and Co., for example, made "Trueste" and "Tiffany" as sprays of EAU DE TOILETTE for use by the ladies. An enamel example which does not have a dot over the letter I is illustrated (fig. 9.98).

Fig 9.98

Fig 9.99a

Fig 9.99b

Fig 9.99c

9.176 EAU DE TOILETTE VERVEINE (Perfume and Toiletry)

A silver slot label with this title was made in 1905 by Britton, Gould & Co. of 83 Hatton Garden, a firm which specialised in making novelties, for Messrs. J. Floris, purveyors of fine perfumes and toiletries to the Court of St. James from around 1730 onwards at 89 Jermyn Street at its junction with Duke street. The stiff paper title slotted into the silver holder. It could easily be changed for a newly acquired perfume or toiletry. The lettering was in three styles. The design was art nouveau, similar to the slot pewter labels with ivory slot titles probably also made for customers of Messrs. J. Floris. See below for VERVEINE.

9.177 Eau de Vie (Medicinal)

Eau de Vie or Aqua Vitae is a colourless fruit BRANDY where the fruit is fermented and then double distilled. It is used as a post-prandial aid to digestion. Stow's Survey of 1754 warned against poor distillation. This

Fig 9.100

typical boudoir label is shown in fig. 9.100 as a decorative unmarked enamel. See also below, GULDEWATER.

9.178 EAU DE VIOLETTE (Perfume)

This title appears on an enamel label illustrated in fig. 9.101. See below VIOLETTE

Fig 9.101

9.179 EAU DENTIFRICE (Toiletry)

See above DENTIFRICE

9.180 EAU DENTRIFICE (Toiletry)

See above DENTIFRICE

9.181 EAU FROIDE (Soft Drink)

This title appears on a narrow rectangular ceramic label with shaped ends decorated with a single reed. The illustration (fig. 9.102) shows it being tied by a blue ribbon probably to a cold water decanter. Originally the label would have been suspended by a chain.

Fig 9.102

9.182 EAU MINERALES BL (Soft Drink)

This was a bin label for use in the store or perhaps a cellar.

9.183 EAU MINERALES N (Soft Drink)

This was a bin label for use in the store or perhaps a cellar.

9.184 EAUX MINERALES (Soft Drink)

This was a bin label for use in the store or perhaps a cellar.

9.185 ELDER (Medicinal, Perfume and Soft Drink)

See ELDER FLOWER below and CURRANT above. ELDER was a popular sauce. ELDER could have been short for ELDERBERRY WINE which was a popular tipple. A pottery bin label with rounded shoulders for ELDER and a bin label for ELDER. FLOWER have been noted. These were probably for ELDER FLOWER WINE but may have been for ELDER. Medicinally liquids with ELDER in were drunk by those with bronchial problems. ELDER oil was certainly used medicinally. It was even used for slimming down. "The Ladies Dictionary: Being a General Entertainment for the Fair Sex", printed in 1694, contains the following recipe for a slimming lotion : "Take an ounce and a half of oyl of foxes, oyl of lilies and capon's grease…adding an ounce of oil of ELDER". Several examples in silver are illustrated (one by Richard Binley with 1756 lion in fig. 9.103, a neck-ring sauce label – by way of comparison- by Jonathan Tayleur circa 1790 in fig. 9.104, a Phipps and Robinson double-reeded rounded – end rectangular in fig. 9.104b, an unmarked plain provincial octagonal in fig. 9.104a and an early Margaret Binley (see fig. 9.104d for maker's mark) curved sauce label with decorative edging in fig. 9.104c) but many of these would have been used as sauce labels. However the octagonal unmarked electroplated example with a beaded edge is in set with ORANGE, CURRANT, RAISIN, CHERRY and GOOSEBERRY and therefore would have been used in the boudoir (see figs. 11.2 & 11.3).

Fig 9.104

Fig 9.103

Fig 9.104a

Fig 9.104b

Fig 9.104c   *Fig 9.104d*

9.186. ELDER FLOWER (Medicinal, Perfume and Soft Drink)

This was a perfume made from the flowers of the Elder Tree. The title appears on an enamel label illustrated as figure 1063 in the book on "Wine Labels" and on a square-shaped silver label in the V & A collection by Charles Rawlings. Confusingly there was also a home-made wine known as ELDER FLOWER for which a recipe is given in the Journal and a soft drink Elderflower Presse is made by Belvoir fruit farms (fig. 9.105a).

Fig 9.105a

Fig 9.105b   *Fig 9.105c*

9.187 Elderflower (Medicinal, Perfume and Soft Drink)

This title in lower case script lettering appears upon a porcelain label mentioned in the Master List, Second Update. See above, ELDER FLOWER.

9.188 Elder flower (Medicinal, Perfume and Soft Drink)

This title has been noted. See above, ELDER FLOWER.

Fig 9.106

Fig 9.107   *Fig 9.107a*   *Fig 9.107b*   *Fig 9.107c*

9.189 ELDER. FLOWER (Medicinal and Soft Drink)

This title appears on a bin label, to ensure easy identification when in storage, and so would not be for a perfume but rather for a drink. See ELDER FLOWER above.

9.190 ELDER FLOWER WATE (Perfume)

This title is shown upon a small French enamel label. See below ELDER FLOWER WATER.

9.191 ELDER FLOWER WATER (Perfume)

Elder-flower water was distilled from the flowers of the elder deciduous shrub and small tree. It was known as acqua sambuci (47). See also ELDER FLOWER and ELDR.FLOR.WATER. A small silver label is illustrated in fig. 9.107 together with a selection of enamels (figs. 9.106, 9.107a-c).

9.192 ELDER-FLOWER WATER (Perfume)

This title has been noted on an enamel label.

9.193 ELDer FLOwer WATer (Perfume)

See above ELDER-FLOWER WATER

9.194 ELDR. FLOR. WATER (Perfume)

This is a delicate silver gilt label clearly made for use in a perfume box by Charles Rawlings and William Summers in 1830 of octagonal shape (48).

9.195 ELDr FLOr WATER (Perfume)

See above ELDR. FLOR. WATER

9.196 ELDERBERRY (Soft Drink)

A silver-gilt boudoir label (fig. 9.108), part of a set of five, was made in 1956 (see CRAB APPLE above). ELDERBERRY was a drink made from the fruit of the tree sambucus nigra. Around 1780 Richard Richardson of Chester made an oval silver label with thread edge affixed to a wire-work loop entitled EBULUM. This label was in set with BEER, MOUNTAIN and WHITE WINE. EBULUM is said to be ELDERBERRY black ale! Elder berries were used for making a strong wine. So most ELDERBERRY labels are wine labels.

Fig 9.108

9.197 ELDER VINEGAR (Medicinal and Toiletry)

A narrow rectangular unmarked double reeded silver label with cut corners and a tiny chain (fig. 9.109) may have been used in the toilet box as a restorative, but its usual use was as a sauce label for a flavoured vinegar in the soy frame.

9.198 ELIXIR (Medicinal, Perfume and Toiletry)

Various ELIXIRS were used to sweeten the breath. They were often yellow in colour and normally aromatised. ELIXIR was a perfume made by Jacques Guerlain and sold in Paris from 1923. The name was derived from the Arabic al-iksir, which literally meant a powder for drying wounds and thus handy to have in one's campaign chest, or perhaps from the Arabic alacsir, by which the alchemists denoted their special philosophers stone and recipe for long life. Dr. Scott in his section on essences explains that the word ELIXIR is indeed of Arabian origin meaning an essence or pure mass without any dregs. The great ELIXIR was of course the one which would turn base metals into gold – the so called philosophers stone or red tincture. An ELIXIR was thought to restore youth and thus prolong life. In medical terms an ELIXIR as supplied by a pharmacist was a powerfully strong extract or tincture. Elixir of Vitriol was a mixture of sulphuric acid, cinnamon, ginger, alcohol and a small quantity of oil of vitriol. An ELIXIR was sometimes a solution of bitter and aromatic vegetable substances in spirits of wine. It was one of a set of eight unmarked silver labels: see ELIXIR DE SPA below. In Donizetti's opera "L'Elisir D'Amore" (the elixir of love) the quack Dr. Dulcamara sold his beauty treatments, cures for ailments and love potions. One medicament allegedly made the poor rich and another was Queen Isolde's ELIXIR which had the potent effects afforded by drinking a home made wine! Daffy's ELIXIR, in either Dicey's version or in Swinton's version, was a compound tincture of senna. In fig. 9.110 a recipe is given for Stoughton's ELIXIR.

9.199 ELIXIR DENTIFRICE (Toiletry)

An enamel label with this title of escutcheon shape and measuring 5.6cm by3.8cm is numbered 422 in the Marshall Collection. It is in set with EAU BORIQUEE and EAU DE TOILETTE. See further DENTIFRICE above.

To make Stoughton's *Elixir*, thin, Are off the rinds of fix *Sevil* oranges V ounce and put them in a quart bottle, with worth ntia fcraped and fliced, and fix pen andy ochineal; put to it a pint of the beft day, e it together two or three times the clear then let it ftand to fettle two days, oonful into bottles for ufe; take a large tea in

Fig 9.110

Fig 9.109   *Fig 9.111*

9.200 ELIXIR DENTRIFICE (Toiletry)

ELIXIR DENTRIFICE is a French enamel label. See DENTIFRICE above.

9.201 ELIXIR DE SPA (Medicinal, Perfume and Toiletry)

This title appeared in Major Gray's 1958 list. It was engraved on an unmarked silver label, one of a set of eight, which also included ELIXIR. So ELIXIR DE SPA was something different. They were oval in shape and thus fall into the period 1775 to 1784. See further ELIXIR above.

9.202 ESCALOTTE (Soft Drink)

Thought to have been an appetiser. There is a restaurant called "L'Eschalote".

9.203 ESCHALOTTE (Soft Drink)

Rawlings and Summers made a silver label with this title in 1832. See above ESCALOTTE.

9.204 ESPRIT DE ROSE (Perfume)

ESPRIT DE ROSE is the name on a small silver-gilt label made by Thomas and James Phipps in 1820 (fig. 9.111). Esprit de Fleurs was a perfume retailed by Guerlain from 1828. See further ATTAR OF ROSES above.

9.205 ESS BOUGET (Perfume)

See ESSENCE BOUQUET below

9.206. Ess. Bouquet (Perfume)

This 1860's label is in set with Lavender Water and Jockey Club contained in a three-bottle slim Tunbridge Ware perfume bottle box veneered in rosewood. See ESSENCE BOUQUET below.

9.207   ESS. BOUQUET (Perfume)
This is the title of a small leaf shaped silver unmarked label (fig. 9.112) in set with OPPOPONAX and LAVENDER. See ESSENCE BOUQUET below.

9.208   ESSENCE (Perfume)
This title has been recorded on a small label made by Susanna Barker in 1791. Paper labels have been found on bottles in a 1630s Augsburg travelling box. These indicate that ESSENCE could stand for "Cinnamon Essence" or "Essence of Balsam" or "Essence of Myrhh". These paper labels were in set with ROSE WATER, WOUND WATER and Spirit of Sweet BALM. Other possibilities include ESSENCE BOUQUET, ESSENCE OF GINGER and ESSENCE OF ROSE.

9.209   ESSENCE BOUQUET (Perfume)
This perfume was made by Guerlain from 1828 in Paris.

9.210   ESSENCE GINGER (Medicinal)
Charles Rawlings made a double reeded pointed oval with this title in 1821 measuring 3.0cm by 1.6cm and numbered 390 in the Marshall Collection. See ESSENCE OF GINGER below.

9.211   ESSENCE OF GINGER (Medicinal)
GINGER is the rhizome of the tropical plant Zingiber Officiniale, characterised by its hot taste. Ginger has certain medicinal properties. A paper label for ESSENCE OF JAMAICA GINGER was included in the Belfast Medical box dating to the 1860s. Essence of ginger was sometimes added to a drink to spice it up a little. ESSENCE OF GINGER must be distinguished from GINGER CORDIAL which is a liqueur. It is one of six unmarked but probably silver labels in the Marshall Collection numbered 346, being in company with POISON LAUDANUM (347), EAU D'ALIBURGH (345), CAMPHtd SPIRIT (344), LAVENDER (348) and TOOTH MIXTURE (349).

9.212   ESS GINGER (Medicinal)
This title appears on a silver label made around 1830 in London by Reily and Storer. This rounded ends rectangular has a maker's mark only and double reeded borders. It measures 2.6cm by 1.4cm and is numbered 389 in the Marshall Collection. It could be in set with a similar label titled CAMPH DROPS. See further ESSENCE OF GINGER above.

Fig 9.112

Fig 9.113

9.213   ESSENCE OF ROSE (Perfume)
Made by Guerlain since 1828, this popular perfume appears in a three-bottle perfume set along with EAU DE COLOGNE and LAVENDER. The silver-gilt labels were made by John Reily around 1815 of rectangular shape with rounded corners. The titles had pierced lettering.

9.214   EYE (Medicinal and Toiletry)
Presumably an abbreviation of EYE LOTION.

9.215   EYE DROPS (Medicinal and Toiletry)
This is the title of a Staffordshire escutcheon enamel in the Vintners' Hall collection. The white enamel is decorated with red roses in each corner of a triangular shape. It is a pair with MOUTH WASH.

9.216   EYE LOTION (Medicinal and Toiletry)
See BORACIC LOTION above.

9.217   EYE WASH (Medicinal and Toiletry)
This title was included in Major Gray's list in 1958. It related to a small late Victorian or Edwardian Crown Staffordshire porcelain travelling toilet box label which has been illustrated in the Journal (49). Similar size porcelain labels have been noted for GLYCERINE and BORACIC. Larger size labels have been noted for BORACIC LOTION and GLYCERINE.

9.218   Eyes (Perfume and Toiletry)
This title appears on a small silver heart-shaped label (fig. 9.113) with decorative border made by Levi and Salaman in Birmingham in 1901 (50) which probably adorned a bottle on a dressing table. This firm was well known for making small silver items, such as a hedgehog pin cushion of 1904 (51), often with a novelty aspect. Eyes is illustrated with ALCOHOL and SHERRY to indicate its small size.

9.219   F.CREAM (Toiletry)

Presumably this stands for face cream, probably contained in a glass jar kept on the dressing table. F. CREAM is a small, unmarked silver label, of rectangular shape with a feathered edge.

9.220   FINE NAPOLEON (Medicinal and Perfume)

FINE NAPOLEON may have been drunk in the boudoir as a restorative like BRANDY. The title appears on a French enamel of around 1890 of broad kidney shape decorated with a border

Fig 9.114

of red and yellow flowers suitable for a Lady's boudoir (fig. 9.114). But in appearance it is very similar to PEROXIDE, which is marked "France" on its reverse side. So it may refer to an EAU DE COLOGNE of this name. Apparently EAU DE COLOGNE was a great favourite of Napoleon. This would explain the decoration and possible pairing with PEROXIDE.

9.221   FINENAPOLEON (Medicinal and Perfume)

This one word title appears on a semi-circular French escutcheon decorated with flowers and blue strokes and hence very suitable for the boudoir. It is in pair with VIELLEFINE, also one word, and is noted in the Master List Second Update.

9.222   FL. D'ORANGE (Perfume)

This title appears on a crescent shaped French enamel. See FLEUR D'ORANGE below.

9.223   FLEUR D'ORANGE (Perfume)

This is a perfume made out of orange flowers. See also EAU d'ORANGE, ORANGE FLOWER, HUILE D'ORANGE, RATAFIAT DE FLEUR D'ORANGE, RATAFIA and RATAFIAT. Fr.d'ORANGE appears upon an oblong continental label of 19th Century date . EAU d'ORANGE appears on a French enamel label. ORANGE FLOWER appears in Major Gray's 1958 list and is a small enamel label. It may have been made as part of a set of three with amhonia and rosewater. It appears as Orange Water in the 13 bottle cordial case as a paper label around 1740. As a constituent of Honey Water it has been said to impart a charming light perfume and to be excellent

for smoothing the skin. HUILE D'ORANGE is the title of a French enamel of around 1770 with floral decoration reportedly seen in Blois. RATAFIAT DE FLEUR D'ORANGE is a French porcelain label once in the Bernard Watney collection. It has also been noted as the title of a Bayeaux porcelain. RATIFIA is a crescent shape silver label made by Phipps and Robinson in 1800 in London, measuring 2.4cm. by 1.1cm. RATAFIAT is a porcelain label. Possibly the best example of FLEUR D'ORANGE is in silver-gilt made by Rawlings and Summers in 1837 (fig. 9.115). Confusingly Fleur d'Orange is also a French term for Orange Muscat which is a wine made from the Muscat grape.

Fig 9.115

9.224   Fleurs d'Orange (Perfume)

This title appears in script lettering on a shield shaped French porcelain. See above FLEUR D'ORANGE.

9.225   Fr. d'ORANGE (Perfume)

See above FLEUR D'ORANGE. The title appears on a Nineteenth Century continental oblong label.

9.226   Floral Lilac Toilet Water (Perfume)

No details available.

9.227   FLORIDA WATER (Perfume)

It appeared in Penzer's Table III. It is a perfume similar to EAU DE COLOGNE in purpose but not equal in popularity in the United Kingdom although it was very popular in America. A silver plated label with this title is numbered 55 in the Marshall Collection and measures 5.6 cm x 3.1 cm. It was a mixture of EAU DE COLOGNE with oil of CLOVES, cassia and lemongrass. A detailed recipe for it is given in The Household Cyclopedia of Perfumery.

9.228   Florida Water (Perfume)

See above FLORIDA WATER.

9.229   FLORIS VERVEINE (Perfume)

See VERBENA below.

9.229a FONTANIAC (Medicinal)

Thomas Edwards made a silver die-stamped label with this pierced title in 1828 of wine label size but with a somewhat feminine floral bouquet design incorporating the rose of England, the thistle of Scotland and the shamrock of Ireland. FONTANIAC is probably a variant spelling of FONTINIAC. The label was produced in a hurry because the title was incised with the label in an upside-down position. It should be distinguished from FRONTIGNAC and FRONTIGNAN which were Muscat wines also taken for medicinal purposes. FONTANIAC (like FONTINIAC) was a cordial used to help with problems of digestion and advertised for that purpose. Interestingly Thomas Edwards had made three years earlier a PEPPERMINT silver label also for use for medicinal purposes.

9.230 FONTENAY (Perfume)

This title appears on a silver plated label.

9.231 FONTINIAC (Medicinal)

FONTINIAC is an unusual title. FONTINIAC is in set with CRAB APPLE, MEAD, ELDERBERRY and PARSLEY, all made by Wakely and Wheeler in 1956, retailed

Fig 9.116

by Garrard and Company, for the best of home-made productions. It is thus in set with four other silver-gilt boudoir labels which includes CRAB APPLE on which see above. The name however appears on Mr. Tyers' list of refreshments available for purchase at six shillings a bottle in the Guide to Vauxhall Gardens published in 1760. The Pavilion was open from 5pm Mondays to Saturdays from the beginning of May until the end of August. So its is thought to have been a medicinally refreshing drink made from crushed grapes of many varieties and colours. It also appears on an unmarked beaded crescent shaped plated label circa 1775 measuring 4.9cm by 1.2cm and numbered 56 in the Marshall Collection. A variant spelling of FONTINIAC is FRONTINIAC, on which see below, paragraph 9.242. On the other hand it could be a mis-spelling

of FRONTIGNAN, which according to Webster's Revised Unabridged Dictionary of 1913 was the name given to a sweet muscadine (Muscat) dessert wine made near Frontignan, a town in Languedoc in Southern France, and along the coast near Montpelier, which is also spelt FRONTENAC, FRONTIGNAC, FRONTIGNAN and possibly FRONTIGNIA. The name FRONTINIAC appears on a small mother-of-pearl crescent label seemingly in set with four wine titles but of boudoir appearance, on an unmarked silver label in set with MADEIRA and RAISIN (52) and on a silver label attributed to Darby Kehoe of Dublin around 1771 to 1780, numbered 62 in the Marshall Collection, which is 5.5cm by 2.8cm in size. FRONTIGIAC appears upon a pottery bin label.

9.232 FOUR BORO' (Perfume)

This title appeared in Major Gray's 1958 list. It has been noted in the Journal as having been in the same ownership as LAVENDER WATER. So perhaps it is related to it in some way.

9.233 FRANCIPANNI (Perfume)

See FRANGIPANI below.

9.234 FRANGIPANI (Perfume)

A popular glove perfume used in Rome had this title. It was said to have been invented by Count Frangipani. Two earlier types of glove perfumes were described by Gervase Markham in 1631 in his "English House Wife". Reference however to FRANGIPANI was made in Act III of The Virtuoso by Thomas Shadwell, performed at Dorset Garden in May 1676, in the context of toilet-waters such as Amber, Orangery, Genoa, Romane, Neroli, Tuberose, Jessimine

Fig 9.117

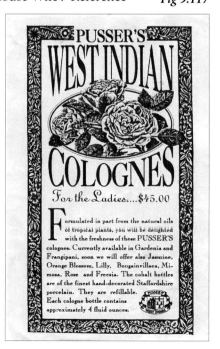

and Marshal. Reference to Frangipan was made by John and Mary Evelyn in 1690 in Mundus Muliebris in the context of Martial (the name of Louis XIV's perfumier), Jonquil, Tuberose, Orange, Violet, Narcissus, Jassemin and Ambrett. The glove perfume was based on spices and orris root with a small amount of musk. In Regency times a perfume made by Guerlain of Paris from 1828, and retailed as "Extract of Frangipani", was distilled from the flowers of the Plumeria Rubra, the red jasmine tree, getting its name from Mercutio Frangipani, the Count's grandson, who apparently discovered this tree in Antigua having accompanied Christopher Columbus to the West Indies. From mid Victorian times it was fashionable to keep trying out new perfumes. So scent bottles in the Boudoir needed continuous updating. For this reason silver slot labels were used in France of which four types are illustrated in the Chapter on Designs, made to a standard of .950 by makers such as Armand Gross and Christofle in Paris around 1870. The design of Armand Gross is used for a pair of silver gilt unmarked labels for FRANGIPANI and for BERGAMOT.

9.235 FRANGIPANNI (Perfume)

See above, FRANGIPANI

Fig 9.117a

9.236 FRANNGIPANI (Perfume)
See above, FRANGIPANI

9.237 FRENCH (Perfume)
A porcelain label (fig. 9.118) marked "Made in England" and "Crown Staffordshire" is of domed kidney shape with gold titling on a white background and measuring 3.6cm by 2.5cm. It is suitable for use on the dressing table to show off a perfume made in France. A tiny triple-reeded scallop-cornered label (fig. 9.118a) was made by Susanna Barker in 1793. Larger size labels for FRENCH exist (see fig. 9.118b). An oval shape label engraved for FRENCH was made by Stokes and Ireland of Birmingham hallmarked in Chester in 1924 with a gadrooned border which may have graced a larger sized perfume bottle. A long Colonial looking unmarked label

Fig 9.118

Fig 9.118b

Fig 9.118a

Fig 9.119

(6cm) has its title engraved for FRENCH supported by two tropical birds, presumably to identify a French wine. The large art deco label stands for FRENCH VERMOUTH, a pair with ITALIAN, which may have been boudoir tipples.

9.238 French (Perfume)
See above FRENCH.

9.239 French Vinegar (Medicinal and Toiletry)
Hamilton and Inches made a pair of kidney shape plain silver labels engraved in script for French Vinegar and Chilli Vinegar in Edinburgh in 1912 in the style of toddy labels (fig. 9.119). Whilst because of the pairing these are probably sauce labels, Chilly Vinegar and French Vinegar may have been used in the travelling box as restoratives. See below VINEGAR.

9.240 FRENCH VINEGAR (Medicinal and Toiletry)
See French Vinegar above.

9.241 FRONTIGNAC (Medicinal)
The Marshall Collection has a FRONTIGNAC by R.W. Smith of Dublin circa 1818 (numbered 59). This silver label is large, measuring 7.0 x 5.2cm.

Sweet dessert wines were popular in the boudoir. See further FONTINIAC above, paragraph 9.231, and FRONTINIAC below, paragraph 9.242.

9.241a FRONTIGNAN (Medicinal)
The Marshall Collection has a FRONTIGNAN by Charles Dalgleish of Edinburgh, 1817 (numbered 60). The maker's mark has been partly overstamped by an unidentified maker. This silver label measures 5.8 x 2.2cm. See further FONTINIAC above, paragraph 9.231, and FRONTINIAC below, paragraph 9.242.

9.241b FRONTIGNIA (Medicinal)
The Marshall Collection has a FRONTIGNIA by John Reily of London, 1813 (numbered 61). The lettering is picked out by niello infilling. This silver label measures 5.2 x 3.2cm. See further FONTINIAC above, paragraph 9.231 and FRONTINIAC below, paragraph 9.242.

9.242 FRONTINIAC (Medicinal)
FRONTINIAC appears upon an unmarked label, which may have had a connection with Richard Richardson of Chester, in set with MADEIRA and RAISIN (53) which were drunk in the boudoir. A crescent shaped example exists made of mother-of-pearl with engraved leaf decoration appropriate for the boudoir. This is in set with MADERA, W. PORT, CLARET and HOCK. See above, FONTINIAC, for further details, and fig. 9.120 below.

9.243 FRUIT (Soft Drink)
Fruit juice was a favourite soft drink tipple in the boudoir. William Twemlow of Chester made a silver label with this title in 1799, as did Richard Ferris of Exeter in set with GINGER of rectangular shape and with double reeded borders by way of decoration in 1796.

Fig 9.120

Fig 9.121

Fig 9.122

Fig 9.123

9.244 GARGLE (Medicinal and Toiletry)
Use of the GARGLE is mentioned in Elizabeth Bennion's "Antique Medical Instruments". The enamel label with this title is of good size showing that it adorned a reasonably large bottle of MOUTH WASH or GLYCO-THYMOLINE, being particularly useful in dealing with sore throats. It has been found in set with ALMOND CREAM, on which see above, ALCOHOL, BORACIC ACID and TOILET WATER.

9.245 GARUS (Medicinal)
GARUS was a CORDIAL designed to deal with stomach disorders according to the recipe of its namesake, Dr. Garus, a Seventeenth Century Dutch physician. The unmarked silver label (fig. 9.122) is quite large and of unusual design (54).

9.246 G. BEER (Soft Drink)
Standing for Ginger Beer, this title appears on an enamel label in the V & A Museum's Cropper collection (55). Fentimans have brewed Ginger Beer since 1905 (fig. 9.123).

9.246a G BRANDY (Medicinal)
See A BRANDY and C BRANDY above. The G stands for ginger.

9.247 Gentiane (Medicinal)
Gentiane, being the French for gentian, is the title of a label made in France. It is a tonic used to aid digestion, derived from flowers of plants belonging to the gentian family. For this reason it was also used as a flavouring for bitters and not surprisingly is paired with Orgeat (see below).

9.248   GIGGLEWATER (Medicinal)

GIGGLEWATER is a large (approximately 7.7cm. x 5.2cm.) enamelled tinware oval label with a cheap looking chain and blue lettering reminiscent of the scripted CHLOROFORM which was a large porcelain oval label.

Fig 9.124

These may have originated in America. GIGGLEWATER could have been a night-time medicinal drink to settle the stomach or could have been, it has been suggested, a coded reference to some sort of alcoholic drink during the times of prohibition, on which an article published in The Times has been reviewed above in connection with Crème de Thé. Prohibition was established by the 18th Amendment to the United States constitution and the Volstead Act which came into force in 1920 banning the sale of intoxicating liquors. It was repealed by the 21st Amendment in 1933, which provided for the regulation of sales to be at State level. GIGGLEWATER appears to be datable to the period 1920 to 1933.

9.249   GINGER (Medicinal and Soft Drink)

GINGER oil is a natural essential oil. GINGER syrup (Syrapus Zingiber) was prepared from the root of Zingibe Officinale. For centuries GINGER has been used to relieve nausea and other digestive problems. It was thought in 1847 to be good for dealing with flatulence, dyspepsia, gout, debility and torpor (56). "King's Ginger" was a favourite tipple of King Edward VII (1901-1910). Ginger was thought to be an aphrodisiac. GINGER may also have stood for GINGER WINE (although bin labels and wine labels are known (57) bearing the title GINGER WINE), which was alcoholic (Peter Ustinov is said to have enjoyed GINGER WINE as a hot toddy at Ye Olde Red Cow at the approaches to Smithfield market), keeping company with WHISKY, DAMSON CORDIAL, COOKING BRANDY and HOLLANDS, or for GINGER BRANDY (see below), keeping company with BENEDICTINE, MARASHINO and CHERRY BRANDY, all measuring 3.25cm by 2.2cm and numbered 239 to 242 in the Marshall Collection and made by Henry Thornton in London in 1895, or for GINGER CORDIAL (see below). Presumably the bin label marked GINGER stands for GINGER WINE (58). It is in set with GOOSEBERRY, PORT

and BRONTE. The Weed Collection included a flat ivory hoop label engraved for GINGER. The Cuthbert Collection included a plain curved rounded end rectangular silver label for GINGER made by David Manson of Dundee around 1810. William Howden of Edinburgh around 1805 made a plain rectangular silver label for GINGER with a thick double reeded border, measuring 4.8cm by 2.0cm and numbered 64 in the Marshall Collection. Richard Ferris made a GINGER label in pair with FRUIT (which see above). A potential boudoir label is the GINGER with an unusual highly decorative design executed by Joseph Glenny around 1819 with a length of 4.7cm. (59). More recently a Coronation label was made for GINGER in pair with CHERRY (see CHERRY above). Larger size GINGER labels may have adorned highly decorated lidded display jars such as were made in the 1930s in the Rouge Royale range by Wiltshire and Robinson. Illustrated are a feathered crescent marked PA (fig. 9.126c), a double-reeded rounded-end rectangular marked M&F (fig. 9.126a), a die-pressed escutcheon in Victorian Gothic design by George Unite made in 1853 (fig. 9.125) and a bone collar (fig. 9.126b).

Fig 9.125

Fig 9.126b

Fig 9.126

Fig 9.126c

Fig 9.126a

9.250 GINGER BRANDY (Medicinal)
GINGER BRANDY is a household preparation keeping company with CLOVES and PEPPERMINT. The title was engraved on one of a trio of large oval labels for stored home-made preparations (fig. 9.127). It was also, however a liqueur, in set with CHERRY BRANDY and other silver labels made in 1895 by Henry Thornton (see above GINGER).

Fig 9.127

9.251 GINGER CORDIAL (Soft Drink)
See below, GINGER-CORDIAL. This title appears on an octagonal plated label with triple reeded borders in Sheffield City Museum (see fig. 9.128).

9.252 GINGER-CORDIAL (Soft Drink)
See above, GINGER CORDIAL. This title appears on an octagonal plated label with beaded borders, measuring 4.9cm. by 2.9cm. and numbered 64 in the Marshall Collection.

9.252a Ginger Wine (Medicinal)
See above C. PORT for explanation and GINGER for further information.

9.253 GINGERETTE (Medicinal and Soft Drink)
This title has been recorded (see fig. 9.129). The label is unmarked.

9.254 GLICERINE (Toiletry)
This enamel label (fig.9.130) probably belongs to a group of four and thus be in company with TOILET POWDER, BATH SALTS and EAU DE VIOLETTE. See GLYCERINE below.

9.255 GLYCERINE (Toiletry)
The hydrate of glyceryl is a triatomic alcohol. In late Victorian times it was a collective name for the group of alcohols to which it belonged. It also gave its name to any preparation consisting of a specified substance dissolved or suspended in GLYCERINE. What was bottled and transported in the travelling toilet box was a colourless, sweet, syrupy liquid obtained from animal and vegetable oils and fats. Sometimes it was mixed with honey, as shown by a bottle labelled for Gille Pharmacie of Paris retrieved from the wreck of the steamship "Minnehaha" which sank off the Scillies in 1910. The modern name for this toiletry preparation is glycerol. The title appears on a late Victorian or Edwardian Crown Staffordshire porcelain travelling toilet box label (fig. 9.131), marked ENGLAND AD1801 pattern number A 6297, possibly a pair with EYE WASH (60). See also below ROSEWATER & GLYCERINE.

9.256 GLYCO-THYMOLINE (Medicinal)
This title appears on a plated label in the Clemson-Young Collection. See above GARGLE. It was probably in set with PONDS EXTRACT, LYSTERINE, BORACIC WATER and CITRATE MAGNESIA. Glyco is derived from the Greek word for sweet. When used in the name of a chemical compound it indicates the presence of glycerol (see GLYCERINE above). Thymol is the phenol of cymene $C10H12OH$ obtained from the oil of thyme. GLYCO-THYMOLINE is thus a powerful antiseptic used as a GARGLE from mid-Victorian times.

9.257 GOLD WATER (Medicinal and Soft Drink)
See below GULDEWATER.

9.258 Gomme (Medicinal)
This title refers to a classic pick-me-up made of gum syrup. It appears on a long banner shaped silver label. The title is in lower case lettering.

Fig 9.128

Fig 9.129

Fig 9.130

Fig 9.131

9.259 GOOSBERRY (Soft Drink)
See below GOOSEBERRY.

9.260 GOOSBERY (Soft Drink)
See below GOOSEBERRY.

9.261 GOOSEBERRIE (Soft Drink)
See below GOOSEBERRY.

9.262 GOOSEBERRIES (Soft Drink)
This title appears on an Old Sheffield Plate label of gadrooned octagonal design. See below GOOSEBERRY.

9.263 GOOSEBERRY (Soft Drink)
An example in Old Sheffield Plate is in pair with DAMSON. An example in silver made by Elizabeth Morley in 1811 is also in pair with DAMSON. Another example is dated 1814 and made by Cocks and Bettridge in Birmingham. It measures 4.3cm by 1.5cm and is numbered 67 in the Marshall Collection. Gooseberries were also used to make a home-made wine. That this was probably stored is indicated by a bin label for GOOSEBERRY made around 1810 (61), in set with GINGER, PORT and BRONTE. Fig. 9.132 shows a plated example and fig. 9.132a shows a triple-reeded octagonal made by Phipps, Robinson and Phipps in 1815.

Fig 9.132   *Fig 9.132a*

9.264 GOOSEBERY (Soft Drink)
See above GOOSEBERRY.

9.265 GOOSERERRY (Soft Drink)
See above GOOSEBERRY.

9.266 GOSEBERRY (Soft Drink)
See above GOOSEBERRY.

9.267 GOUTTES de MALTHE (Medicinal)
Known in English as Tar Water, this preparation was good for alleviating chronic bronchitis. An example formed part of the H.E. Rhodes collection and the name was mentioned in Major Gray's list of 1958. A French porcelain label with this title is in the V & A

Museum collection. Wood tar was a popular remedy as an expectorant to deal with sub-acute and chronic bronchitis. Apparently one part of tar water was dropped into four parts of water. Coal tar, the source of benzene, is a compound of many substances. In its preparation it had to be strained and filtered. Coal tar was also applied externally as a skin disease lotion.

9.268 GOUTTES DE MALTHES (Medicinal)
See above GOUTTES de MALTHE.

9.269 GRAIN (Medicinal)
See Graine de lin below.

9.270 Graine de lin (Medicinal)
A French porcelain shield shaped label bears the title Graine de lin in script, standing for linseed oil. It matches Vinaigre ordinaire.

9.271 Grandad's Medicine Bottle (Medicinal)
A medium sized oval electro-plated nickel silver label, made in the 1910s by Yeoman of England (1897-1972) in Sawbridgworth, Hertfordshire, bears this intriguing script title Grandad's Medicine Bottle. The label is bordered in the traditional manner with a single reed and gadrooning. The chain is silver. Yeoman Silver Plate supplied high quality EPNS to Asprey's, Fortnum's, Harrod's and Liberty's to name only a few of its important retail customers.

9.272 GRANDE MAISON (Perfume)
This was the title used on one of the set of eight oval silver labels acquired by Mr. R.B.C Ryall. It was in the company of other titles which were clearly perfumes and so formed part of a perfume set of labels. Perhaps it was for use on a big house occasion. See further above ACQUA D'ORO.

Fig 9.133   *Fig 9.134*

9.273 GRAPE (Soft Drink)
Squeezed grape juice was very refreshing for the boudoir. Its container could have been marked with a small Mother-of-Pearl label. Samuel Knight made

a silver label for GRAPE in 1816 and Reily and Storer did likewise in 1839 of rectangular shape with cut corners, triple reeded borders and pierced lettering, measuring 4.0cm by 2.5cm and numbered 69 in the Marshall Collection. Illustrated (fig. 9.133) is a Phipps, Robinson and Phipps example of 1816 with its owner's initials "GSB" on the reverse.

9.274 GRAPE FRUIT (Soft Drink)
See GRAPEFRUIT below.

9.275 GRAPEFRUIT (Soft Drink)
Grapefruit juice was a popular refreshment. The much appreciated GRAPEFRUIT cordial was known as the seductive "Forbidden Fruit". The title appears on a plain octagonal single reeded silver label by Charles Kay (1815-1827) in the Ryall Collection (62).

9.276 GREEN (Soft Drink and Writing Ink)
The title GREEN has been recorded on a label mentioned in the Master List. It may refer to a silver label in set with or similar to the writing ink label COPYING (which see above), or to a label for GREEN Tea or GREEN Tea and Orange (fig.

Fig 9.135

9.135) made with some 32% Green Tea and other added flavourings mainly Orange Juice but along with Apple Pomace, Blackberry Leaves, Hibiscus, Cinnamon, Citric Acid, Malic Acid, Ginger, Orange Peel and Clove. Samuel Palmer, the so-called English Van Gogh, and his artistic friends the "Extollagers" long after the last "amber glow of the neighbours' candlelit lattices had faded" would be sitting in the Water House at Shoreham in Kent "indulging in their agreeable ceremonies of pipe-smoking and snuff-taking" as they sipped away at their bowls of "dear precious GREEN Tea". This drink was their particular favourite and even years later a bowl of it would stir nostalgic memories for Palmer of quiet evenings in Shoreham and "nice old-fashioned talks by the fire". (63).

9.277 GREENGAGE (Soft Drink)
GREENGAGE is a fruit juice cordial. A label with this title is in set with APRICOT (which see above)

and others.

9.278 GUJAVA (Soft Drink)
This is a variant spelling for Guava, a fruit originating in the West Indies from which it was common practice to produce a juice (64). The name appears upon a Danish silver label of around 1830 made by Nicolai Christensen.

9.279 GULDEWATER (Medicinal and Soft Drink)
Gould water was a natural mineral water pumped up to ground level. It took its name from that of the Gould pumps used to pump up the water. It was sometimes called GOLD WATER, gulden water or GULDEWATER. Gould water was one of the thirteen titles of cordial bottles contained in a Dutch travelling box of around 1740. A bin label with this title would have helped one locate mineral water in storage. GOLD WATER or GULDEWATER was also the name given to a cordial medicine, sometimes described as a herbal liqueur, recommended for dealing with heart problems and for the palliation of leprosy. A recipe for making it was given in Latin in a medical treatise written by Arnold de Villeneuve, revealing what he thought had been a best kept secret for preserving youth and retarding the onset of old age. It consisted of an Eau de Vie to which one added Rosemary Flowers, Cinnamon, Cloves and Liquorice. The Rosemary Flowers gave it a golden colour – hence the name. Where Eau de Vie de Dantzig was used as a base some minute quantities of gold dust were added to make the product seem like a kind of ELIXIR long cherished by pharmacists.

9.280 GUM (Medicinal)
See below GUM SYRUP.

9.281 GUM SYRUP (Medicinal)
An oblong gadrooned silver label (fig. 9.136) made by Charles Tiffany of New York (1891-1902) bears this title. From early times it was a remedy for rheumatic complaints. A recipe given by W. Buchan in 1785 recommended that three or four

Fig 9.136

ounces of GUM (sometimes called guaicum) should be infused in a bottle of rum or brandy. A later recipe of 1946 shows that it was, according to Charles H. Baker Jr., a gentleman's pick-me-up.

9.282   H (Perfume, Toiletry)

This small initial label is likely to stand for Huile which was in much demand. However it could stand for HUNGARY WATER as the label is similar to and possibly in set with the initial label engraved LW for LAVENDER WATER. The tiny size of both labels makes it likely that they were for use on perfume bottles in a perfume or toilet set. It was made by Joseph Willmore of Birmingham in 1811 (65).

9.283   HAIR LOTION (Toiletry)

The title is known from a published list of titles of labels. For hair lotions see further EAU DE PORTUGAL above and YLANG – YLANG below.

9.284   HAIR TONIC (Toiletry)

Dr. Scott gives a recipe for an "Oil for thickening the hair" in his Toilette No. V (66), which is illustrated along with a silver HAIR WATER label in fig. 9.138. An American embossed paper label, the product of F.E. Mason & Sons of Batavia in New York State dateable to May 1906, advertises PRINCESS HAIR TONIC made by a firm in Richmond Virginia based on Eau de Quinine. Another advertises CROWN HAIR TONIC and scalp cleanser, stated to be for external use only and to be based upon an alcoholic content of 25%. This preparation was made by J.B. Lynas & Son, perfumers. The title is well suited for an enamel label (fig. 9.137).

Fig 9.137

Fig 9.138

9.285   HAIR WASH (Toiletry)

This title has been recorded

9.286   HAIR WATER (Toiletry)

This title appears on a silver label made by Thomas Diller in 1832 (fig. 9.138).

9.286a   HALLONSAFT (Soft Drink)

A continental silver label with this title for CHERRY CORDIAL was part of Lot 316 in Christie's sale of 5th November 2002. It was not illustrated. It was in set with CITRONSAFT (LEMON JUICE) and KORSBARSSAFT (RASPBERRY AND REDCURRANT CORDIAL).

9.287   HAMMAM BOUQUET (Perfume)

This title appears on a large Bilston enamel of around 1875 (fig. 9.139). It was retailed for sale by Penhaligon of London's Covent Garden from 1872 onwards. An embossed paper label with this title in a red shield format has been noted attached to a 3.25 inches high perfume scent bottle (also shown in fig. 9.139). In the late 1860s William Henry Penhaligon quit Penzance for London. He established himself first as a barber and then as a supplier of pomades and toilet waters and finally as a perfumer, creating his first rich and floral fragrance in 1872, which he called Hammam Bouquet after the Turkish Baths which had perhaps inspired the creation. He created other fragrances for members of the Royal Family and the nobility. Royal Warrants were awarded to Penhaligon by Queen Alexandra in 1903, the Duke of Edinburgh in 1956 and the Prince of Wales in 1988. He created Blenheim Bouquet in 1902 (one of the first citrus scents to be created) for the Duke of Marlborough, which indeed was used by Sir Winston Churchill.

Fig 9.139

9.288   HARTSHORN (Medicinal)

Spirits of Hartshorn (see below "Hartshorn") were derived from shavings of the horn or antler of the hart, a male red deer, by destructive distillation, and were used in Dr. Scott's recipe for a liniment for baldness. Solutions of ammonia in water (liquid ammonia) and ammonium bicarbonate were also called HARTSHORN and used as a cure for a sore throat (see above GARGLE and GLYCO-THERMOLINE) according to John Wesley in his "Primitive Physic" published in 1747. HARTSHORN was incidentally also used as a leavening or raising agent in baking before the introduction of baking soda.

9.289 Hartshorn (Medicinal)
This script title has been found on a double reeded octagonal silver label (fig. 9.140) made in 1822 by William Fountain (also attributed by some to William Frisbee but he ceased working in 1820). See above HARTSHORN.

Fig 9.140

9.290 HAY (Perfume)
HAY is believed to have been a fragrance rather like the smell of newly mown hay. A domed escutcheon enamel label, formerly in the Lank Collection, bears this title, decorated with flowers above and what appear to be black grapes below, but none-the-less joyfully enamelled in shades of pink, green, yellow, purple and red, on a white background, typical of a boudoir label (fig. 9.141). See below NEW MOWN HAY.

Fig 9.141

9.291 HONEY AND ALMOND CREAM (Toiletry)
This perhaps was used as a hand lotion, to keep skin soft and supple. Honey Water, which imparted a delightful light perfume, was sometimes mixed with almond cream See above, ALMOND CREAM.

9.292 HUILE (Medicinal, Perfume, Toiletry)
HUILE is a pair with VINAIGRE. The French language is used to reflect the status of France as regards perfume manufacture by the English silversmith Herbert Charles Lambert in 1908. The labels are made in silver-gilt which would be appropriate for perfumes. The choice of enclosed crescents for a design however suggests that they may have been made for an oil and vinegar frame. Equally, they could have adorned two glass bottles in a travelling box for use in hot weather.

9.293 HUILE. D (Medicinal)
This title, presumably a shortened form of HUILE

D'ANIS either DES INDES or ROUGE, appears on a South Staffordshire enamel label of circa 1790. Oil d'anis, otherwise Pimpinella Anisum, was regarded as good to be taken to deal with indigestion.

9.294 HUILE. D'ANIS DES INDES (Medicinal and Perfume)
See above ANIS. The title appears on an English enamel label of circa 1790 (fig. 9.142). The East India Company imported anise, peppermint, spearmint, lime, lemon grass, citronella and citrodora. The apostrophe is rather faint.

Fig 9.142

9.295 HUILE D'ANIS ROUGE (Medicinal and Perfume)
Oil of red anise is made from the umbellate flowers of pimpinella. See above ANIS. The title appears on a French enamel label, belonging to "Family 36" according to the Circle's classification drawing upon the research conducted by Dr. Richard Wells, the smallest size pre-Samson French label group of around 1770. So there is a dot over each capital letter "I" in the title. It is in set with CRÈME DE PORTUGALE and EAU. D'OR, on which see above.

Fig 9.143

9.296 HUILE D'ORANGE (Perfume)
The perfume label bearing this title is said in the Journal to be "an English enamel decorated with flowers, circa 1770, seen at Blois, France" (67).

9.297 HUILE de ROSE (Perfume)
Oil of Rose is a well known essential oil.

9.298 Hui.le de Rose (Perfume)
Fig 9.144

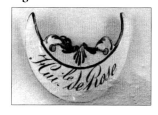

The title is shown by black lettering on a white dome-shaped French enamel crescent label (fig. 9.144). See above HUILE de ROSE.

9.299   HUILE DE VANILLE (Perfume)
This title appears on a French enamel escutcheon belonging to Family 30 (middle size), keeping company with CHIPRE(see above). It also appears on a silver plated oval shaped label with a double reeded border, which is similar in style to a label of 1820 for EAU-DE-PORTUGAL, and on a rounded ends rectangular once in the Ryall Collection illustrated in the Journal (68). HUILLE DE VANILLE was one of Mr. Ryall's eight oval silver labels in the ACQUA D'ORO set. It was produced as a perfume or as a lesser strength toilet water.

9.300   HUILE DE VENUS (Medicinal and Perfume)
The title HUILE DE VINUS appears on an oval silver label, unmarked, a pair with EAU DE COLODON. It was used in a set of six Eighteenth Century glass perfume bottles assembled in a travelling case (69). The title HUILE DE VENUS has been recorded as a Battersea label paired with MUSCAT (see above, fig. 1.2). It was an exotic perfume for men as well as an oil having curative and soothing powers as a vegetal lubricant for sexual comfort which is still produced in France (fig. 9.145). It comprises the oils of Amandes and Jojoba and the essential oils of Rose and Geranium.

Fig 9.145

9.301   HUILE DE VINUS (Medicinal and Perfume)
See above HUILE DE VENUS.

9.302   HUILE DE ANIS (Medicinal and Perfume)
This title appears as No. 19 in Penzer's Table III.

9.303   HUILE DE ROYAUX (Perfume)
This title appears as No. 20 in Penzer's Table III. Thomas Halford made a large label, some 5.8cm by 2.0cm, with this title circa 1807 (maker's mark TH only) of crescent shape with gadrooned borders which is cased (so that the reverse side is not viewable) and numbered 570 in the Marshall Collection.

9.304   HUILLE DE VANILLE (Perfume)
See above HUILE DE VANILLE

Fig 9.146

Fig 9.147

9.305   HUILLE DE VENIS (Medicinal and Perfume)
See above HUILE DE VENUS.

9.306   HUILLE DE VENUS (Medicinal and Perfume)
See above HUILE DE VENUS.

9.307   HUNGARY (Medicinal and Toiletry)
See below HUNGARY WATER

9.308   HUNGARY WATER (Medicinal and Toiletry)
Hungary Water, an astringent lotion, was an ingredient of Milk of Roses. An example was in the H.E. Rhodes Collection (70). Another example is a silver-gilt escutcheon made in 1826 by Charles Rawlings in the V & A Museum's Cropper collection (fig. 9.146). The title was included by Major Gray in his 1958 list and by Dr. Penzer in his list of Toilet Waters. An example made in 1835 by Joseph Willmore of Birmingham is also illustrated (fig. 9.147). According to one account it consisted of rectified alcohol, rosemary oil, lemon peel oil, oil of balm, oil of mint, ESPRIT DE ROSE and FLEUR D'ORANGE. According to another it was prepared from various herbs and especially included rosemary flowers. According to a 1999 recipe given in "Herbs for Natural Beauty"(71) for use as an astringent lotion, one should add six parts of lemon balm, four parts of chamomile, one part of rosemary, three parts of calendula, four parts of roses, one part of lemon peel, one part of sage and three parts of comfrey leaf to a full bottle of vinegar adding (optionally) the essential oil of either lavender or rose. It is made to-day by Crabtree & Evelyn. It is a very old recipe said to have acquired its name from Queen Elizabeth of Hungary for whose Court it was made around 1370. There are references to it in 1698, 1706 and 1747 where it is described as "Queen of Hungary's Water". In fact John Wesley gave two "prescriptions" for it in 1747. By reputation this herbal product was

used as a kind of cure-all to treat gout and lameness, as a hair rinse, a mouth wash, a cure for headaches when mixed with white-wine, vinegar and water, as an after-shave lotion, as an astringent lotion and to treat sore feet in a foot-bath. See also ROSEMARY and ROSEWATER.

9.309 INDIAN (Soft Drink)

See below, INDIAN TEA. The enamel label (fig. 9.148) is small but it may however refer to Indian Tonic Water!

Fig 9.148

9.310 INDIAN TEA (Soft Drink)

This label was possibly made for a Ladies Travelling Box for a person who liked to take her favourite tea leaves with her for a restorative brew. This could have been very soothing after a bumpy carriage ride. It could have adorned a tea caddy or canister in the boudoir. The Marshall Collection has, numbered 430, a Staffordshire bone china label of escutcheon shape measuring 3.8cm by 2.6cm with this title which would probably have hung on a displayed container in the boudoir.

9.311 Infusion d'Homme (Perfume and Toiletry)

This is a generic title for men's fragrances including Cologne, Eau de Toilette, Aftershave Lotion, Perfumed Bath Soap and in more modern times Shower Gel. All these fragrances are available to-day from Prada in Milan.

9.312 IODINE (Medicinal)

IODINE, mentioned in Major Gray's 1958 list, was applied to cuts and bruises. It was obtained from seaweed. This label may have been a pair with WATER.

9.313 JAM. VINEGAR (Medicinal)

Probably standing for JAMAICA VINEGAR this variety of VINEGAR is thought to have served a medicinal purpose. The title appears on a small, oval silver label, typical of one used in a Toilet Box. It was presumably a restorative.

9.314 JASMIN (Perfume)

This perfume was retailed by Jacques Guerlain in Paris from 1928, having been prepared in Grasse. A label with this title is in the Cropper Collection in the V & A Museum. A JASMIN bearing the inscription

"Dreyfous 99 Mount Street W." is recorded in the Journal (72). See also below JASMINE and JESSAMINE.

Fig 9.149

9.315 JASMINE (Perfume)

This fragrant perfume made by the maceration process became popular from the 1860s. So it is the title appearing on a middle size French enamel which is possibly a pair with EAU DE ROSE. These labels bear the retailers' mark of Dreyfous of of 99 Mount Street in the West End of London. The character of the style of lettering used is very idiosyncratic. The letter S has seraphs. The letters J and I have a dot above. A similar style is used for much smaller labels such as WHITE LILAC and METHYLATED SPIRITS. JASMINE Sambac and JASMINE Graniflora are two types of natural essential oils. See also above JASMIN and below JESSAMINE.

9.316 JESSAMINE (Perfume)

This title appears as No. 21 in Penzer's Table III. It is a perfume as old as FRANGIPANI, ORANGE FLOWER and NEROLI. It is mentioned as Jessimine in Thomas Shadwell's "The Virtuoso" published in 1676. John Reily made a silver-gilt label with this title around 1820, a pair with PORTUGAL (which see below). It is numbered 575 in the Marshall Collection, where it is suggested in the catalogue that JESSAMINE is a variant spelling of JASMINE and is a toilet water often displayed upon a Ladies Dressing Table.

9.317 Jockey Club (Perfume)

This 1860's label is in set with Lavender Water and Ess. Bouquet contained in a slim Tunbridge Ware perfume bottle box veneered in rosewood. Jockey Club Bouquet was produced from the 1850s by Daniel Rotely Harris of D.R. Harris & Co. of 55 (now 29) St. James's Street in the West End of London.

9.318 JONQUIL (Perfume)

This title appears in Major Gray's 1958 list. It may have been part of a set with JASMIN and VIOLET in the Cropper Collection in the V & A Museum. Two or three heads of the strongly scented JONQUIL (Narcissus Jonquilla) or rush-leaved daffodil (from the Latin juncus, a rush) will affect a room in no unmistakable way.

9.319 Kersevaser (Medicinal and Soft Drink)
See below, KIRCHEN WASSER.

9.320 KHOOSH (Medicinal)
The Chanticleer Society of cocktail enthusiasts has a bottle with a paper label for KHOOSH Bitters. Its contents were golden in colour and tasted like dark rum. The label referred to the medicinal value of the bitters which were guaranteed to contain no harmful substances. Another bottle of KHOOSH bitters is in the Museum of The American Cocktail. Fig. 9.149a is unmarked except for "Made in England".

Fig 9.149a

9.321 KIRCCH WASSER (Medicinal and Soft Drink)
See below KIRCHEN WASSER.

9.322 KIRCHEN WASSER (Medicinal and Soft Drink)
This drink is a fruit cordial made from the wild black cherry of the Black Forest in Germany which gives it its German name. It was described in the "Lay of St. Romwold" from the Ingoldsby Legends published around 1840 as a "rich juice". Flavourings for cordials used in the boudoir were certainly gathered from many parts of the world (73). As in the case of NOYAU and ANISSETTE there was an alcoholic version of this cordial supplied by Plasket & Co. of Old Burlington Street in London around 1860 to Captain C.R. Egerton RN (74). His estate realised £885 for his cellar contents in 1870. Joseph Taylor of Birmingham made a label with this title in 1824. It had a wine-label type moulded vine border (75). A boudoir version in silver-gilt was made by Paul Storr. It appears to be a pair with NOYAU. Both labels are marked for 1835. A larger version has been noted (5.75cm wide) also in a rococo heavy cast scrolling cartouche style (76). The labels were titled KIRSCHENWASSER. A tiny label giving yet another version of the name this time for Kersevaser in lower case lettering is noted in the Master List Second

Update, made of white porcelain with gilt scrolling around the eyelets which is suggestive of boudoir use.

9.323 KIRCHEVASER (Medicinal and Soft Drink)
This title has been recorded in the Journal (77). See above KIRCHEN WASSER.

9.324 KIROCHWASSER (Medicinal and Soft Drink)
This is the title upon a mother-of-pearl label (78). See above KIRCHEN WASSER.

9.325 KIRSCH A WASSER (Medicinal and Soft Drink)
See above KIRCHEN WASSER.

9.326 KIRSCHENWASER (Medicinal and Soft Drink)
This title has been recorded in the Journal (79). See above KIRCHEN WASSER.

9.327 KIRSCHENWASSER (Medicinal and Soft Drink)
See above KIRCHEN WASSER. KIRSCHENWASSER is to-day produced in Switzerland as a commercial liqueur. An enlargement of Paul Storr's silver-gilt label is illustrated to to demonstrate its quality (fig. 9.150).

Fig 9.150

9.328 KIRSCHWASSER (Medicinal and Soft Drink)
Around 1830 in London Reily and Storer made a large 5.4cm x 3.5cm. silver label with this title in pierced lettering. As a boudoir label it had a floral and foliage border. See above KIRCHEN WASSER.

9.329 KIRSCH-WASSER (Medicinal and Soft Drink)
See above KIRCHEN WASSER

9.330 KIRSHWASSER (Medicinal and Soft Drink)
In 1824 in Birmingham Joseph Taylor made a silver label with a similar title. See above KIRCHEN WASSER.

9.330a KORSBARSSAFT (Soft Drink)
A continental silver label with this title for RASPBERRY AND REDCURRANT CORDIAL was part of Lot 316 in Christie's sale of 5th November 2002. It was not illustrated. It was in set with CITRONSAFT (LEMON JUICE) and HALLONSAFT (CHERRY CORDIAL).

9.331 LA (Perfume)
This title has been taken from the Master List. No particulars are given in the Journal. It is thought to be a corruption of LAVENDER, which see below.

9.332 LA ROSE (Perfume)
A large silver label made by William Summers in 1871 is of gorget shape and pierced for LA ROSE. It has been illustrated in the Journal where the maker is said to be William Smiley (80). LA ROSE was a favourite Eau de Parfum in Victorian times used in quantity which might explain the size of the label. Fragrances available to-day include Lavin's "Jeanne La Rose", Mugler's "La Rose Angel", Coty's "La Rose Jacqueminot" and Goutel's "La Rose". Farrow and Jackson have produced a bin label for LAROSE, numbered 452 in the Marshall Collection, which may possibly relate to the storage of LA ROSE, but more likely of Rose Water for the rose water bowl or sprinkler.

9.333 LACHRYMA (Medicinal)
The seeds of the herb coicis lachryma-jobi are used to make a liquid medicinal preparation to counter inflammation. It sometimes referred to as Job's tears. This title is easily confused with wine-titles Lachrymae Christi or Lachryma Christi, sometimes referred to as Christ's tears, and their various versions which confusingly include LACHRYMA. The Tears of Christ is the name given to celebrated sweet red and dry pale golden yellow Neapolitan wines produced from grapes grown on the Southern slopes of mount Vesuvius in the Campania district of Italy. Label size and decoration is the best guide to purpose. An attractive plated example is in the Anderson Collection housed in Vintners' Hall (fig. 9.151a) and a small silver label is also illustrated (fig. 9.151b).

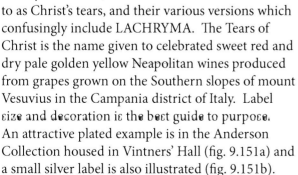

Fig 9.151a

Fig 9.151b

9.334 LACKRYMA (Medicinal)
See LACHRYMA above.

9.335 LACRIMA (Medicinal)
See LACHRYMA above.

9.336 LACRYMA (Medicinal)
See LACHRYMA above.

9.336a LAIT D'IRIS (Perfume)
The title is believed to have appeared on a Crown Staffordshire porcelain label produced around 1900 and retailed by Thomas Goode & Co. for use on a perfume bottle. The perfume was admired as being a great aid to seduction as well as being used by both sexes.

9.337 LAROSE (Perfume)
See LA ROSE above.

9.338 LAUDANUM (Medicinal)
This was at one time sold by the pharmacist as a tincture of opium. It was obtainable without any controls at the date of the Medical Box found by Mrs Marshall which she "came upon with all its bottles and labels intact". It was however known to be a poison and would normally be carried by the doctor in the secure compartment in the back of his specially designed travelling medical box. Poisons were marked as such. POISON LAUDANUM is a title recorded by Penzer in his Table III at No. 27. A label with this title made of silver or silver plate, unmarked, a double reeded rounded rectangular measuring 2.6cm x 1.4cm, is in the Marshall collection. Farley's recipe for Liquid Laudanum is reproduced in the text. A paper label for LAUDENUM with the lettering POISONOUS in red is contained in the Belfast medical box of around 1860 and indicates a variant spelling.

9.339 LAVANDE (Perfume)
Two French enamel labels with this title have been noted: one of broad triangular shape with a floral design in a predominantly blue colour (fig. 9.152a) and one in a small size of kidney shape with floral sprays and a blue border at the bottom (fig. 9.152b). These perfumes were probably produced in Grasse, famous for its essential oils.

Fig 9.152a

Fig 9.152b

9.340   LAVENDAR (Perfume)
LAVENDAR is a silver splayed neck-ring hoop label, being part of a set of eight with ATTAR OF ROSES. See LAVENDER below.

9.341   LAVENDER (Perfume)
LAVENDER is a natural essential oil, is an effective cleanser and has healing and soothing properties. It refreshes, relaxes, revives and renews and is therefore used in aromatherapy. It has a delicious scent. This title has been noted on an octagonal label with a floral border, on a modern 1960's label and on a silver-gilt shell and scroll label by John Reily circa 1820 and on a label by Phipps, Robinson and Phipps made in 1812 and similar to but not a pair with ROSE of the same design and date but different in size. It was a very popular perfume. It appears in the silver-gilt perfume boxed set made by John Reily around 1825, now in the Marshall Collection, numbered 334 to 336, along with ESSENCE OF ROSE and EAU DE COLOGNE. LAVENDER, in company with BOUQUET and OPPOPONAX, is the title on one of a 1930s set of three art deco unmarked silver labels (fig. 9.158). Alcool Pur de Lavande was sold in Paris in the 1820s. In London from about 1790 onwards Harris' Apothecary in St. James's Street established a reputation in selling quality Lavender Water. To-day one can buy from D.R. Harris & Co. Ltd., by appointment to H.R.H. The Prince of Wales, Chemists, LAVENDER soaps, bath oils, essences, hand lotions, body lotions, shaving soaps and shaving creams. "The Hermit in London", published in 1819, records this instruction from a Guards Officer to his servant : "Let me have the last boots which Hoby made for me, not the Wellingtons nor the iron-heeled ones, but the last ones with copper heels; and be sure to use the blacking which has marasquina in it and oil of lavender". See also LA, LAVENDAR, LAVENDER WATER, LW, EAU DE LAVANDE, LAVANDE, and SPIRIT LAVENDER.

Fig 9.153

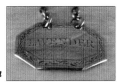

Fig 9.153a

Fig 9.153b

Fig 9.154

9.342   LAVENDER WATER (Perfume)
This name, included in Penzer's Table III, has been noted on many labels (see figs. 9.155 to 9.157) including (i) an octagonal silver-plated example with a floral border; (ii) a silver vine-leaf by Rawlings and Summers in 1831 (fig. 9.155); (iii) a shaped silver label by Rawlings and Summers in 1836; (iv) a heart-shaped porcelain label (fig.9.156); (v) an enamel label decorated with a blue ribbon; (vi) a silver hoop or collar label engraved "EVL" made around 1830 (81); (vii) two Staffordshire English enamels; (viii) an unmarked Staffordshire type porcelain label with blue riband, deep red roses and dark green foliage (fig. 9.157); (ix) a silver label of octagonal shape without decoration by William Summers in 1887, with EAU DE COLOGNE as a pair; (x) a silver label with AMMONIA and ROSEWATER in set; (xi) a silver label with FOUR BORO' and METHYLATED SPIRIT in set; and (xii) an enamelled escutcheon label with blue borders in the Vintners' Hall Collection. EAU DE LAVANDE was one of a pair with ROSE GERANIUM. Raffald in 1799 gives a recipe for distilling LAVENDER WATER which is reproduced in the text. Also illustrated is a paper label evoking the spirit of Simon Bolivar in 1830 (fig.9.157a). Several monasteries, convents and specialist lavender farms in Europe produce lavender oil and lavender water. Illustrated (figs. 5.10 and 5.11) are two scenes taken from the archives of "The Times" taken in August 1957 showing the Cistercian monks of Caldey Island, off the coast of Pembrokeshire, picking lavender to produce Caldey Island Toilet Water, and one of the monks measuring the ingredients of a Caldey Island perfume. A LAVENDER WATER was in Major Gray's 1958 list, and in the Bignall Collection (82).

9.343   Lavender Water (Perfume)
See above LAVENDER WATER. This lower case lettering title is in set with Ess. Bouquet and Jockey Club.

9.344   L'EAU CHAREC (Medicinal)
This was a kind of Ratafia pick-me-up or cordial having, according to Mew and Ashton's "The Cordial and Liqueur Maker's Guide" of 1894, a basis of white wine, popular from 1780 to around 1820. An entry for September 23rd in the 1801 Farington Diary explains its use in Scotland. "After dinner was removed but before the fruit course was served the ladies retired to the withdrawing room. A case

Fig 9.155 *Fig 9.155a* *Fig 9.155b* *Fig 9.155c* *Fig 9.156* *Fig 9.156a*

Fig 9.156b *Fig 9.156c* *Fig 9.156d* *Fig 9.156e* *Fig 9.156f* *Fig 9.157*

Fig 9.157a

Fig 9.158

of small decanters was placed before the host's wife, who helped her guests to very small glasses of various stimulants as they liked. This is a custom and the ladies partake of it." L'EAU CHAREC is often found in set with LEMON (perhaps for Lemon Brandy), BURGUNDY (perhaps for Burgundy Cordial), ELDER (perhaps for Elder Wine) and CHERRY (perhaps for Cherry Brandy).

9.345   LEDSAM VALE (Medicinal)
This was a refreshing cordial in vogue in 1829.

9.346   LEMON (Medicinal and Soft Drink)
LEMON oil is a natural essential oil. A small oval ribbed silver label with the title LEMON was made by William Ellerby between 1810 and 1845. There are many other examples. Eight examples in silver are illustrated: see fig. 9.159 for an unmarked beaded crescent; fig. 9.160 for a Rawlings and Summers' neck-ring of 1829; fig. 9.161 for a double-reeded octagonal by James Hyde in 1796; fig. 9.161a for a double-reeded octagonal by Elizabeth Morley circa 1795; fig. 161b for a double-reeded rounded-end rectangular by Joseph Angell in 1829; and figs. 161c-e for escutcheons with pierced titles all probably by John Reily circa 1825. Boudoir labels would probably be around 2.6cm by 1.3cm. but it is not possible to be definitive. Whilst LEMON was quite often used on a soy frame bottle it could well have hung on a decanter containing lemon cordial as a restorative drink, or could have hung on a medicine bottle containing lemon, honey and ginger to contain a common cold and cure a sore throat, or could have been hung around a tea caddy. LEMON and GINGER tea was a great reviver and revitaliser. Such labels would have been decorated, not just reeded or left plain. LEMON appears upon a small decorative urn-shaped label made by Phipps and Robinson in London around 1795. Limoen Water was one of the thirteen titles appearing on cordial bottles in a Dutch travelling case of around 1740 in date. LEMON majestically appears on a scroll below an earl's coronet below a crested dragon. The label, shown in fig. 9.161f, is unmarked.

Fig 9.159

Fig 9.160

Fig 9.161

Fig 9.161a

Fig 9.161b

Fig 9.161c

Fig 9.161d

Fig 9.161e

9.348   LEMON BRANDY (Medicinal)
See above BRANDY.

9.349   L. JUICE (Soft Drink)
See above LEMON and below LIME JUICE. The title is used on a large plated oval shaped label with a beaded border, suspended by the use of two foliate projections, for use on a jug of juice, measuring 5.5cm by 3.2cm and numbered 85 in the Marshall Collection.

9.350   LILAC (Perfume)
LILAC is a toilet water prepared from the flowers of syringe vulgaris. The title appears upon a French enamel label fitted with gilt chains in the form of a basket of flowers. It was made by Dreyfous.

Fig 9.161f

9.347   Lemon Balm (Medicinal)
See above BALM. LEMON and GINGER tea produced by Twinings (fig. 9.162) was designed to revive and revitalise.

9.351   Lily of the Valley (Perfume)
This was one of a set of four pewter perfume slot labels, the titles being hand-written and changeable.

Fig 9.163

9.352   LIME JUICE (Soft Drink)
LIME JUICE was one of a set of four reeded rectangular labels made by John Reily around 1801. Belvoir fruit farms make to-day a lime and lemongrass presse (fig. 9.164). It prevented scurvy. English sailors were known by Americans as Limeys (82a).

Fig 9.162

Fig 9.164

9.353 Limonade (Soft Drink)

This title appears on a Limoges porcelain label (fig. 9.165) in pair with Ratafiat de fleur d'orange. The names are scripted. Lemonade would have been very refreshing taken during a journey or at a picnic. This drink was made out of lemons with water and sugar added. The French for lemon is limon. Limonade is a French label although this spelling was used in England according to the Oxford English Dictionary, but its pair is undoubtedly French.

Fig 9.165

9.354 LIMOSIN (Medicinal and Toiletry)

This was a silver-plated label. LIMOSIN was, like its probable pair SPRUCE, a flavouring derived from a tree, in this case the oak, often applied to toiletries. It was at one time in the Ryall Collection.

9.355 Limoux (Perfume)

This French enamel label is a pair with Pt. George. Perfumes have been named after Limoux and Pt. George. However Limoux is a city in Languedoc, which has four recognised wine growing districts. This confuses the issue.

9.356 LIQUID POWDER (Toiletry)

A somewhat unusual Crown Staffordshire porcelain label "made in England" with sloped lettering bears pattern number A6297, painter's mark "F" (fig. 9.166) and was retailed by T. Goode & Co. of South Audley Street in the West End of London. It has a small individualistic chain. Thus it is not chain "A" of the GLYCERINE and BORACIC set, chain "B" of the EAU DE ROSE and EAU DE COLOGNE set or chain "C" of the BATH POWDER design. It would not appear to belong to any set of porcelain labels so far noted. LIQUID POWDER was presumably some sort of BATH POWDER on which see above.

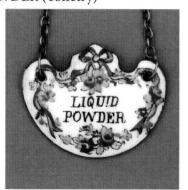

Fig 9.166

9.357 LIQUID SOAP (Toiletry)

This title was in Major Gray's list of 1958. There was no cross reference to details of any label. It was not unusual for guests to take their favourite soap with them when staying at a country house.

9.358 LISTERINE (Toiletry)

This title appears in Penzer's Table III at Number 23. Still popular to-day this label marked out a bottle of mouth wash and gargle of American origin. A plated example was in the Clemson-Young collection.

9.359 LOCH LOMOND (Perfume)

This title appears on a small (3.7cm. by 1.6cm.) mother-of pearl label of crescent shape numbered 94 in the Marshall Collection.. This green perfume has the very familiar scent of coniferous trees in high country.

9.360 LOVAGE (Medicinal)

LOVAGE is a hardy perennial herb belonging to the family of Umbelliferae. As long ago as the time of Pliny it was used as a remedy for sore throats and as an aphrodisiac. It is often drunk with BRANDY.

Fig 9.167

A plated example is in the Anderson Collection, somewhat large but very attractive (fig. 9.167).

9.361 L.W (Perfume)

This small silver unmarked perfume label probably stands for "Lavender Water". It is of the 1810s period but could be earlier. The title is finely engraved. The label measures 2.5cm by 1.6cm. It is similar to the "H" of 1811 by Joseph Willmore which might stand for HUILE or for HUNGARY water.

Fig 9.168

9.362 LYSTERINE (Toiletry)

See above LISTERINE. With this spelling it is said to be a popular American gargle (see GARGLE above) and mouth wash (see MOUTH WASH below). The title appears upon one of a probable set of five plated labels along with CITRATE MAGNESIA, PONDS EXTRACT, GLYCO-THERMALINE and BORACIC WATER.

9.362a MADEHERE (Soft Drink)
A long narrow rectangular silver label inscribed MADEHERE by L.F. Newlands of Glasgow produced around 1800 may relate to a home made cordial.

9.363 MAGNESIUM CITRATE (Medicinal)
MAGNESIUM CITRATE is an oral solution often lemon flavoured, the ingredients of which include citric acid, lemon oil, polyethylene glycol, purified water and sodium, used as a laxative for the relief of occasional constipation or irregularity. It generally produces bowel movement within six hours according to information from a manufacturer. Citrate is a salt of the sharp tasting citric acid found in the juice of oranges, lemons, limes and citrons, the fruit of the tree Citrus medica (see CITRATE MAGNESIA). Magnesium is the element, whereas magnesia is the oxide of the element. CITRATE MAGNESIA is one of a possible set of five silver-plated labels along with LYSTERINE, PONDS EXTRACT, GLYCO-THERMALINE and BORACIC WATER. A paper label for MAGNESIUM CITRATE was included in the Belfast medical box of around 1860 in date.

9.364 MALVERN WATER (Soft Drink)
MALVERN WATER may have been kept in a night decanter by the bedside. An unmarked silver or silver plated example was in the Tubbs Collection. A crescent shaped unmarked silver-plated version with stamped lettering measuring 4.9cm. by 1.0cm. is numbered 105 in the Marshall Collection. Another open crescent shaped unmarked silver-plated version is illustrated. It is fitted with a slip-chain for ease of cleaning. One end of the chain simply has a hook on it so that one end of the chain can easily be unfastened from the eyelet of the label. Yet another crescent shaped version has been reported as having been sold in a saleroom (83).

Fig 9.169

9.365 MAPLE SYRUP (Medicinal)
This silver label, made by SP inLondon in 1987, of oval shape with a cut out stag in long grass, with the name shown below the grass, may have been intended for use in the boudoir.

9.366 MARMALADE (Soft Drink)
A set of three enamels, for PLUM, RASPBERRY and MARMALADE, with red borders, a red rose above and yellow flowers below, has titles in black with distinctive lettering which

Fig 9.170

may refer to fruit juices or cordials kept in decanters. The labels are too large to be used on Edwardian silver jam pots (see fig. 9.170).

9.367 MEAD (Soft Drink)
Whilst mead was well known as an alcoholic liquor made by fermenting a mixture of honey and water, it was also a soft drink beverage charged with carbonic acid gas and flavoured with syrup of sarsaparilla. It was a popular drink in the boudoir from 1890 onwards. The silver-gilt MEAD is part of a five label set which includes CRAB APPLE which see above. A spiced or medicated form of mead was known as Metheglin.

Fig 9.171

9.368 MEADE (Soft Drink)
The rectangular plated label in the Marshall Collection (numbered 110) is thought to be for liquor but it may well be an alternative spelling of MEAD.

9.369 MELOMEL (Soft Drink)
This unmarked label, which has been illustrated in the Journal (84) and was once in the Ryall Collection, made of silver or silver-plated, in narrow rectangular style with a decorated border, was used to identify a cordial made out of sweet blackberries (see CORDIAL)

9.370 METHELATED SPIRIT (Medicinal)
See below, METHYLATED SPIRITS.

9.371 METHYLATED (Medicinal)

This version was engraved on a plain curved silver narrow rectangular rounded to pointed end label made by Thomas H. Hazlewood of Birmingham in 1913 (fig. 9.172) and on a silver label made by Peter and William Bateman in London in 1807 (fig. 9.173). It also appears upon two similar creamy backed English enamel labels   See below, METHYLATED SPIRITS.

Fig 9.172   *Fig 9.173*

9.372 METHYLATED SPIRIT (Medicinal)

A small enamel label with this title made circa 1890 was in the Hollick Collection. Enamel and porcelain labels are illustrated (figs. 9.174a-f) including a small English enamel with green coloured borders perhaps made in Bilston (fig. 9.174g). A silver rectangular label with this title was made by Mappin and Webb in London in 1914 (fig. 9.175). See below, METHYLATED SPIRITS.

Fig 9.174

Fig 9.174b   *Fig 9.174c*

Fig 9.174d   *Fig 9.174e*

Fig 9.174a

Fig 9.174f   *Fig 9.174g*

Fig 9.175

9.373 METHYLATED SPIRITS (Medicinal)

To methylate is to mix or impregnate with methyl. Meths were used to deal with skin complaints and cold sores. Meths were also used to fuel the burner on a kettle stand to provide hot water. Mrs Raffald's recipe for distillation of 1799 is illustrated (fig. 9.176a) along with a small enamel label for METHYLATED SPIRITS possibly used in a travelling box. Spirits of wine mixed with methyl were exempted from duties imposed on alcohol and used for medicinal purposes!

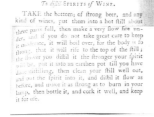

Fig 9.176a

9.374 MILK (Soft Drink)

This title appears on a plated oval shaped label measuring 5.5cm by 3.1cm with a beaded edge and numbered 113 in the Marshall Collection.

9.375 Milk of Roses (Perfume)

This title was recorded by Major Gray in 1958. In 1808 Phipps and Robinson made a small octagonal silver-gilt label with double reeded borders and Milk of Roses in script now in the Cropper Collection (85). It was a toilet water popular during the late eighteenth Century and whole of the nineteenth Century. It consisted of pulverised almonds (see above ALMOND CREAM), mixed with ROSEWATER (see below). Penzer says that after slowly straining the mixture one adds one quarter of an ounce of oil soap together with two ounces of alcohol in which was dissolved half a dram of OTTO OF ROSES (see ATTAR OF ROSES above). Penzer was relying upon a description given by John Timbs in 1868 in editing "Lady Bountiful's Legacy". It was used to beautify the skin and keep it smooth, fair and white looking. In 1826

Fig 9.176b

Dr. Scott gave three recipes for this cosmetic perfume (illustrated in fig. 9.176b). Another recipe for the complexion is given in "The Household Cyclopedia of Perfumery".

9.376   MILK PUNCH (Soft Drink)

The Anderson Collection includes two similar plated labels for MILK PUNCH and TANGERINE (see further below) which were possibly non-alcoholic for use in the boudoir. In Queen Anne's reign MILK PUNCH might have been a restorative made of sugar, milk, herbs and spices. John Reily made a small MILK PUNCH label in 1805 of rectangular shape with cut corners and a triple reeded border, measuring 4.3cm by 2.5cm and numbered 114 in the Marshall Collection. Joseph Willmore of Birmingham made two small labels, one for PUNCH in 1830 and one for MILK PUNCH in 1820 with a decorated border (86). MILK PUNCH was also made in quantity. The Marshall Collection has, numbered 456, a Wedgwood bin label with this title. MILK PUNCH has also been associated with CHARTREUSE GREEN. A silver label by Charles Rawlings in 1823 is sauce label size (87). Another silver label by Frederick Brasted in London in 1875 is smallish (fig. 9.177). It looks like a boudoir label being engraved with flowers and chased with scrolls (88). Alexander Stewart's label of circa 1820 is wine label size. This would have been used for the alcoholic version, for which Elizabeth Moxon in her "English Housewife" (1764) gave a recipe which involved taking "2 quarts of old milk, a quart of good brandy, the juice of six lemons or oranges ….. and about 6 ounces of loaf sugar" and mixing these ingredients together. See also fig. 9.177a.

9.377   MK PUNCH (Soft Drink)

This is thought to be an abbreviation of MILK PUNCH, which see above. This title also appears on bin labels.

9.378   MILLE FLEURS (Perfume)

This name appears in Penzer's Table III as Number 24. See MILLE FLUERS below.

9.379   MILLE FLUERS (Perfume)

MILLE FLUERS is in a set of four perfume bottle labels said to be used for replenishing vinaigrettes. These little silver labels, along with ALBA FLORA and ALBAFLOR, must be among the earliest known, being made by Margaret Binley between

Fig 9.177    Fig 9.177a

Fig 9.178

1764 and 1778. In the same set were AROMATIC VINEGAR, VERBENA and EAU DE COLOGNE. The better view is that these four labels adorned bottles in a perfume toilet box set rather than bottles standing on a shelf within easy reach for use and for replenishment (89).

9.380   MILTON (Medicinal)

This name has been recorded. It is thought to refer to an anticeptic preparation.

9.381   MINT (Medicinal and Perfume)

MINT is an aromatic herb belonging to the family Lamiaceae. MINT was one of the eight titles in the silver perfume set which included CITRON, CLOVES, LAVENDAR, ROSEMARY, WOODBINE and ATTAR OF ROSES. An attractive plated oval rococo cartouche label for MINT , measuring 4.7cm by 3.2cm, is numbered 115 in the Marshall Collection. MINT appears in the second edition of Historia Plantarum, written in Latin by L.U. Grune and published in 1567 in France for use in a medical college. It has Sixteenth century hand-written annotations in brown ink, giving the English names of all the plants illustrated by woodcuts used for medicinal purposes as an aid to easy identification. MINT was used by pharmacists certainly in the late Victorian age as a flavouring. A chemist's bottle, in pair with BITTERS, brightly coloured for display purposes, is illustrated (fig. 9.178). MINT is also used to flavour tea and so could be a tea caddy label.

9.382   MIXT (Perfume)

Standing for Mixed Herbs, this title has appeared in a perfume set, representing mixed fragrances.

9.383   MON PARFUM (Perfume)

Fig 9.179b

This title appears on two enamel labels, one in the V & A Museum and one with an oriental figure displayed (90). This certainly personalised the

choice of toilet water. Tiffany named an Eau de Parfum as MON PARFUM, perhaps with this in mind. The Samson design of an oriental figure, measuring 4.85cm by3.1cm, was also used for EAU DE TOILETTE. To-day MON PARFUM is a CHYPRE fragrance for women launched by Paloma Picasso in 1985. Its top notes are said to be hyacinth, ylang-ylang, bergamot, angelica, rose and lemon (91).

Fig 9.179a

9.384   MOSS ROSE (Perfume)

This is the title of a small enamel label with a ribbon top. The perfume, sometimes called "purslane", was distilled from the flowers of porta laca grandiflora, a plant which grows in Brazil and Argentina.

Fig 9.180

9.385   MOUTH WASH (Toiletry)

This name has not surprisingly appeared on enamel labels, being either French by Samson or English probably South Staffordshire. The name was listed as Number 25 by Penzer in his Table III. It was a common name for GARGLE, which see above. The Staffordshire triangular shaped label is in the Vintners' Hall Collection, a pair with EYE DROPS, of white enamel with red roses in each corner.

Fig 9.181

Fig 9.182

Fig 9.182a

9.386   MUM (Medicinal)

MUM was a good night drink to send one to sleep It has been described as a remarkable liqueur imported first from Germany towards the end of the Seventeenth century. Dean Aldrich from Christ Church Oxford, who was a great fan of unusual catches involving curious wines and spirits (like FLORENCE, PERRY and SHIM) hailed its arrival

with a catch of his own:

> "There's an odd sort of Liquer new come
> from HAMBOROUGH,
> 'Twill stitch a whole Wapentake thorough
> and thorough;
> 'Tis yellow, and likewise as bitter as Gall,
> As strong as six horses, coach and all.
> As I told you, 'twill make you drunk as a
> Drum –
> You'd fain know the name on't, but for that,
> my Friend, MUM!"

Andre Simon in his "English Wines and Cordials" quotes a recipe for it of 1691 which contains, amongst other things, ground beans, shoots of fir and birch, cardamoms, a great variety of herbs and berries and ten new laid eggs. It seems to have been a home made brew from the 1690s. The recipe concludes "Doctor Egidius added water-cress, wild parsley, and six handfuls of horse-radish to each hogshead, and it was observed that the MUM which had in it the horse-radish drank more brisk than that which had not" (92). MUM is also said to be a very strong BEER, taking its name from brewer Christian Mumme. It has been said to be good for hushing clamorous crowds. Beers made with spruce, birch leaves and herbs were a speciality of Brunswick.

9.387   MURON (Perfume)

MURON is a perfume with a base of olive oil infused with a wide variety of aromas, flowers and scents from South West France. As used by the Catholic Church it is a consecrated mixture of olive oil, flowers, herbs and scents. Children are anointed with MURON when they are baptised into the Catholic Church in Armenia.

9.388   MUSCAT (Perfume)

MUSCAT is an overpowering musk perfume popular in the sixteenth and seventeenth Centuries. The title appears on a Battersea enamel label, possibly a pair with HUILE DE VENUS. It

Fig 9.183

has given its name to the Muscat Rose, which is sometimes called a musk rose, because of its musky colour. Larger labels with this title may be wine labels relating to sweet wines made from the Muscat grape, such as MUSCAT DE LUNEL. Eight bottles of MUSCAT were in the Ashburnham sale (93). MUSCAT is the title of an escutcheon shaped enamel label measuring 6.4cm by 4.5cm and numbered 426 in the Marshall Collection. A pottery bin label for MUSCAT has been noted.

9.389   MUSCATE (Perfume)
Josiah Snatt made a small silver label for MUSCATE in London in 1804 for what is assumed to be a variant spelling of MUSCAT. See above, MUSCAT.

9.390   NECTARINE (Soft Drink)
This title appears on a label made by John Emes in 1799 once in the Tubbs Collection. It could relate to a fruit juice.

9.391   NEW MOWN HAY (Perfume)
NEW MOWN HAY is thought to have been a fragrance popular in America capturing that rather special smell of new mown hay in the harvest field. The title appears on a silver crescent shaped label impressed with a .925 silver standard mark, in set with TOILET WATER, probably of the art deco period.

9.392   NOIEAU (Medicinal and Perfume)
See NOYAU and NOYEAU below. This title appears on a mother-of-pearl label perhaps used in a travelling box.

9.393   NOYAU (Medicinal and Perfume)
This boudoir title appears on (i) a small sized ivory label probably used in a travelling toilet box of uncertain age because ivory is difficult to date and was not often used for labels, but it was a popular bottle collar because splayed hoops could be cut

Fig 9.186a   *Fig 9.186b*   *Fig 9.186c*

directly concentrically from the elephant's tusk; (ii) a re-used and re-titled sauce label made by Reily and Storer originally for CAYENNE pepper now re-engraved NOYAU in pair with HARVEY sauce which has been re-engraved COGNAC which is also a restorative; (iii) a double reeded rounded rectangular silver label by George Unite in 1827 measuring 4.3cm by 3.4cm and numbered 120 in the Marshall Collection; (iv) a tiny label, the same size as one for ROSE.WATER, a perfume, measuring 2.9cm by 1.9cm, made around 1830 by Reily and Storer, being as one might expect a highly decorative oval in shape with a single reeded and gadrooned border (fig. 9.184); and (v) two unmarked labels, one in silver (fig. 9.185) and one plated (fig. 9.186). Note the use of the title EAU DE NOYAU – see above – and see also NOIEAU above and below NOYEAU. NOYAU and NOYEAU were both stored, as shown by the existence of bin labels mentioned in the Master List First Update. See also figs. 9.186a-c above.

9.394   NOYEAU (Medicinal and Perfume)
Examples in silver include labels made by John Robins 1796, Thomas and James Phipps 1820, measuring 3.1cm by 2.3cm and numbered 121 in the Marshall Collection, John Reily 1825, Mary and Charles Reily 1826, William Knight 1830 (fig. 9.187), Robert Garrard 1846 in silver-gilt (fig. 9.188) and Rawlings and Summers 1856. Note the small sizes. As well as a perfume and a flavouring, for example for sweet dishes, NOYEAU was a home-made restorative drink. Mrs Beeton gives a recipe for it: one takes 4 pints of milk, 2 pints of whisky or gin, a pound of caster sugar,2 ounces of bitter almonds, one ounce of sweet almonds, the rinds of 3 lemons and 1 tablespoonful of honey and having blanched and

Fig 9.184

Fig 9.186

Fig 9.185

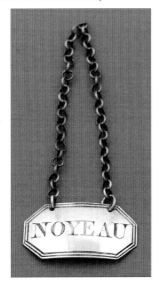

Fig 9.187

Fig 9.188

Fig 9.188a

Fig 9.188b

pounded the almonds and having boiled the milk, mix all these ingredients together. You then allow it to stand until quite cold and for a further 10 days shaking them every day. Finally, you filter the mixture through blotting paper, then you bottle off the mixture for use in small bottles and seal the corks. This recipe makes about 24 pints of NOYEAU. Other similar recipes exist including one of 1825 where the ingredients are the same but the quantities slightly different. A recipe of 1797 refers to a cordial (or pick-me-up) made with French BRANDY, prunes, celery, bitter almonds, a little essence of lemon-peel and ROSE WATER, further flavoured by the kernels of apricots, peaches and nectarines. The average cost of making home-made NOYEAU was about 2 shillings and 9pence at this time. Note the use of the titles EAU DE NOYEAU and NOYAU – see above. NOYEAU, along with KIRSCHEN WASSER and others, was described as a "rich juice" in the Lay of St. Romwold (94). See also figs. 9.188a & 9.188b.

9.395   NOYEO (Medicinal and Soft Drink)

Perhaps standing for a non-alcoholic NOYEAU (see above) because of its connection with RASPBERRY, this title appears on an Old Sheffield Plate label in Sheffield City Museum, a pair with RASPBERRY. However as a grape and vine design appears between the horns of a well opened crescent this pair of labels may well relate to alcoholic medicinal cordials – boudoir tipples in fact – and RASPBERRY could refer to RASPBERRY brandy.

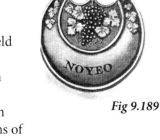

Fig 9.189

9.396   NOYEU (Medicinal and Soft Drink)

See above NOYEAU.

9.396a O D C (Perfume and Toiletry)

A plain small silver label made by John Batson (marked on the face of the label) in 1880 bears the engraved title ODC presumably standing for Eau De Cologne.

Fig 9.189a

9.396b O.D.C. (Perfume and Toiletry)

This title has been recorded and may be another version of ODC for Eau De Cologne.

9.396c O D V (Perfume and Toiletry)

A plain small silver label made by John Batson (marked on the face of the label) in 1880 bears the engraved title ODV presumably standing for Eau De Violette.

Fig 9.189b

9.396d O.D.V. (Perfume and Toiletry)

This title has been recorded and may be another version of ODV for Eau De Violette.

9.396e ODV de Danzic (Perfume and Toiletry)

This title has been recorded and presumably relates to Eau de Violette de Danzic, a toilet water preparation, although it has a modern meaning in economic computation where Optimised Deprival Values are short term valuations and differ from the classical labour theory using Danzig's approach! ODV also stands for Eau de Vie, the potent spirit distilled from fermented fruit juice.

9.397   OEILLET (Toiletry)

This title is the diminuative of oeil meaning eye in French and stands for the eyelet or socket for the eye. It is thought to relate to an eye-wash for dealing with problems around the eye: see above EYE LOTION and EYE WASH.

9.398   OIL (Medicinal, Perfume and Toiletry)

See above HUILE, HUILE D, HUILLE DE ANIS, HUILE D'ANIS DES INDIES, HUILE D'ANIS ROUGE, HUILE de ROSE, HUILE DE VANILLE, HUILE DE VENUS, HUILE D'ORANGE and HUILLE DE ROYEAUX for examples of oils. OIL is a common sauce label being a pair with VINEGAR, usually displayed on a soy frame or an oil and vinegar frame, and used for salads and

Fig 9.190a

Fig 9.193

Fig 9.190   *Fig 9.190b*   *Fig 9.192*   *Fig 9.193a*

flavouring. The boudoir oil would have been an essential oil used for toiletry, medicinal and perfumery purposes. Both boudoir and sauce labels would have been small labels, around 2.5cm by 1.2 cm. Boudoir labels would be more likely to have decoration than sauce labels to show off bottles on the dressing table or in the travelling box. Sauce labels can best be identified by being in set with other soy labels. A possible boudoir label for OIL made by James Newlands and Philip Grierson in Scotland of 1813 has been illustrated in the Journal (95). As regards the labels which are illustrated in figs. 9.190, 9.190a and 9.190b, the first is unmarked, the second is by Yapp and Woodward in 1850 and the third bears Elizabeth Morley's maker's mark only.

9.399 OPPOPONAX (Medicinal and Perfume)

This tiny unmarked leaf – shaped label (illustrated in fig. 9.191) with art deco lettering was included in a perfume set along with ESS BOUQUET and LAVENDER. OPPOPONAX may well have also had medicinal uses. A box drawer in the premises of The Society of Pharmacists of Bruges in Belgium, alongside and forming part of the complex occupied by St. John's Hospital which dates back to at least 1181, for use by pharmacists, is inscribed CHIRON. OPOPAN. and G.RES. OPOPANAX.

Fig 9.191

9.400 ORANGE (Medicinal, Perfume and Soft Drink)

ORANGE oil is an essential oil. Mrs Beecher's Recipe Book (1710-1720) contains a recipe for ORANGE Water which is a perfume used when dining. Elisabeth Morley made a silver label with this title in 1814. George Knight made a silver label with this title in 1820, numbered 125 in the Marshall Collection. It appears upon a plated set of labels along with CHERRY, CURRANT and RAISIN. It also appears upon an enamel label in the Vintners' Hall Collection. ORANGE could be an alternative name for curacao as this drink was based on orange peel and carried its flavour. The silver unmarked ORANGE which is illustrated is in set with PEACH and BENEDICTINE , all labels being suspended by an unusual type of chain with three rings, helps to prove the point being made. ORANGE may also refer to the wine called HERMITAGE because it comes from vineyards near to ORANGE in France. Labels are also known for ORANGE GIN (in silver-gilt measuring only 2.3cm by 1.4cm numbered 126 in the Marshall Collection) and for ORANGE LIQUER (made of silver in 1828 by Reily and Storer of rounded rectangular shape measuring 4.7cm by 2.5cm). See also below, ORANGE BITTERS, ORANGE CURACAO, ORANGE FLOWER, ORANGEADE and POMADE for ORANGE Pommade. Of the labels illustrated, only one is marked, being made by Charles Rawlings in 1820 (fig. 9.193a). Fig. 9.192 is silver and figs. 9.193 and 9.193c are plated. ORANGE tea was also popular (fig. 9.193d)

Fig 9.193b

Fig 9.193c

Fig 9.193d

Fig 9.194

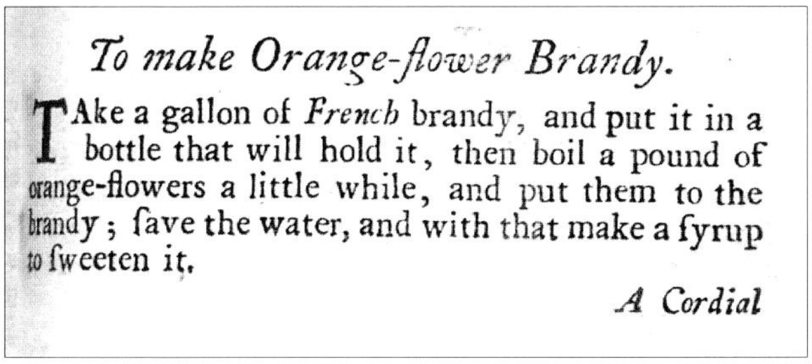

> *To make Orange-flower Brandy.*
>
> Take a gallon of *French* brandy, and put it in a bottle that will hold it, then boil a pound of orange-flowers a little while, and put them to the brandy; save the water, and with that make a fyrup to fweeten it.
>
> *A Cordial*

Fig 9.195

9.401 ORANGE BITTERS (Medicinal)

ORANGE BITTERS were used for medicinal purposes, mainly as a digestive. They were made from the peels of Seville Oranges, cardamon, caraway seed, coriander and burnt sugar, along with sometimes certain herbal additives. Stirrings' version was non alcoholic. A silver label with this title, a pair with ORANGE CURACAO, was made in 1916 by Henry Archer and Company of Sheffield. It measures 3.5cm by 1.8cm and is numbered 263 in the Marshall Collection. Illustrated in fig. 9.194 is a Chester example made by R & S in 1907.

9.402 ORANGE CURACAO (Medicinal)

ORANGE CURACAO was used as a restorative. A silver label with this title in pair with ORANGE BITTERS was made in 1916 by Henry Archer and Company of Sheffield. It measures the same as the label for ORANGE BITTERS and is numbered 264 in the Marshall Collection.

9.403 ORANGE FLOWER (Medicinal and Perfume)

ORANGE FLOWER was the name given to capillaire, a syrup or infusion of maidenhair fern (Adiantum capillus Veneris), designed to deal with heart complaints, pulmonary catarrhs, certain kinds of swellings and even hair loss. Dr. William Kitchener's "Cook's Oracle" and the American "Cook's Own Book" of 1840 both contain interesting recipes. ORANGE FLOWER is the name of a very small enamel, once in the Pratt Collection, used on a perfume bottle. Fleurs d'Orange was the name of an expensive French perfume. It was also a pick-me-up: a recipe (fig. 9.195) for the cordial Orange Flower Brandy advises one to " take a gallon of French brandy, and put it in a bottle that will hold it, then boil a pound of orange-flowers a little while, and put them to the brandy; save the water, and with that make a syrup to sweeten it". Other pick-me-ups were called ORANGE Gin, ORANGE Bitters, ORANGE Curacao and ORANGE Liqueur. Stirrings' Blood Orange Bitters were non-alcoholic. ORANGE FLOWER Water was a constituent of Milk of Roses according to one of Dr. Scott's recipes. And indeed ORANGE FLOWER Water was one of the thirteen titles on cordial bottles contained in a Dutch travelling case made around 1740. ORANGE FLOWER might even have given its name to MARASCHINO but CHERRY would have been the more likely. See also Orgeat below.

9.404 Orangeade (Soft Drink)

This title in script letters appears on a French porcelain shield shape label. It was a popular and refreshing soft drink. Spa water was used with medicinal properties in some cases.

9.405 Orgeat (Medicinal)

Orgeat , the title of a French label, a pair with Gentiane (which see above), was a sweet syrup made from almonds, sugar and either ROSEWATER or ORANGE FLOWER Water and was used for medicinal purposes. Its dominant taste is that of almonds. Thus it was also used as a flavouring.

9.406 PAFAIT AMOUR (Medicinal and Perfume)

See PARFAITE AMOUR below.

9.406a Paniagua (Soft Drink)

This title appears on an unmarked silver escutcheon. The thitle is in script for water. However arguments have been made for it to stand for a little known Spanish wine.

9.407 PARFAIL AMOUR (Medicinal and Perfume)
See PARFAITE AMOUR below. As this title appears upon a pottery bin label being an alternative spelling of PARFAIT AMOUR, it is perhaps unlikely to relate to PARFAITE AMOUR, but it may do so.

9.408 PARFAIT AMOUR (Medicinal)
This liqueur was sometimes drunk for medicinal purposes. See PARFAITE AMOUR below. Illustrated in fig. 9.196 is a small silver-gilt perfume label with this title made by Rawlings and Summers in 1827. According to Dr. Scott in 1826 it was a perfumed liqueur with medicinal properties.

9.409 PARFAIT D'AMOUR (Perfume)
See PARFAIT AMOUR above.

9.410 PARFAITE AMOUR (Perfume)
This Irish perfume label by John Toleken of Cork is a pair with COLARES (see above). It must be distinguished from PARFAIT AMOUR which is the name given to a commercially produced liqueur. According to the Bartender's Guide to the Galaxy it is one of the oldest known liqueurs. It originated in Languedoc.

Fig 9.196

9.411 PARSLEY (Soft Drink)
PARSLEY seed oil is a natural essential oil. PARSLEY is a milky juice made from the aromatic leaves of the biennial umbelliferous plant called petroselinum sativum. The name appears on one of a set of five silver-gilt boudoir labels (fig. 9.197) including CRAB APPLE, which see above for details.

Fig 9.197

9.412 PARSNIP (Medicinal and Soft Drink)
This title may refer to a non-alcoholic version of PARSNIP wine or to the wine itself. The label produced by Susanna Barker circa 1785 for PARSNIP could be for either purpose. It was of rectangular shape with cut corners but it had triple reeding. The alcoholic drink was apparently used to calm the nerves.

9.413 PEACH (Medicinal and Soft Drink)
PEACH Water was one of the thirteen titles on cordial bottles contained in a Dutch travelling case of around 1740. Larger size labels for PEACH may have been for PEACH BITTERS or PEACH BRANDY. A plated octagonal label with a beaded edge for PEACH measures 4.8cm by2.8cm and is numbered137 in the Marshall Collection. The unmarked silver label for PEACH which is illustrated in fig. 9.198, having an unusual chain with three rings in it, is part of a set with ORANGE and BENEDICTINE.

Fig 9.199a

Fig 9.199b

Fig 9.198

9.414 P. MINT (Medicinal and Soft Drink)
Standing for PEPPERMINT (which see below), this title, which appears on an Old Sheffield Plate crescent shaped label, was probably used for some kind of cordial, medicine or tea.

9.415 PEPPERMINT (Medicinal and Soft Drink)
PEPPERMINT oil is an essential oil. Mentha peperita, alias PEPPERMINT, has long been associated with invigorating powers. Mentha was a beautiful Greek nymph. Peperita is a reference to the

Fig 9.200

Fig 9.201

Fig 9.202

9.417 PEROXIDE (Toiletry)

This name is shown on a French Samson enamel. It is similar to another enamel bearing the title of ALCOHOL. PEROXIDE of Hydrogen is used to treat hair by bleaching or dyeing. It formed part of the grooming procedure. See fig. 9.204.

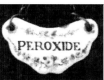

Fig 9.204

spicy pungency of menthol oils. PEPPERMINT was thought to be good for stimulating the appetite. It re-vived and re-vitalised. It also made a very good tea, so important in the boudoir context, to soothe the stomach and create a nice warm glow inside. Thus a small PEPPERMINT label could have been hung on a tea caddy. A label with this title is in the V & A Museum. Another, numbered 141 in the Marshall Collection, was made by James Collins of Birmingham in 1844. It is an attractive oval shaped label set in a rococo foliage frame. It measures however 4.7cm by 3.0cm. Another is broad rectangular in shape with cut corners, made of Old Sheffield Plate, is for a cordial as it is in pair with GINGER CORDIAL (which see above) and is in Sheffield City Museum. These two large labels were presumably used in connection with the enjoyment of PEPPERMINT as a commercially produced liqueur. See also P.MINT above. PIPPERMINT is part of a Chinese set of labels and may relate (see Master List First Update). PEPPERMINT (fig. 9.200) is a large plated label in set with GINGER BRANDY and CLOVES. Pepper was taken to counteract fever. Peppermint tea was popular (see fig. 202). A small medicinal label is illustrated (fig. 201) made by N. Smith & Co. in Sheffield in 1810.

9.416 PERFUME (Perfume)

PERFUME is the title shown on an escutcheon shaped probably English Staffordshire enamel with pink roses, foliage and two small blue flowers (fig. 9.203). With 15 to 40 % perfume concentrates, PERFUME is the purest form of scented product. There is a reference in the Journal to a PERFUME label where it is described as an "enamel escutcheon, red coloured, mid 19th Century" (96).

Fig 9.203

9.418 PICK-ME-UP (Medicinal)

This title appears in Major Gray's list of 1958. In the main it was used as an early morning refreshment after a good party the night before. A range of suitable medicinal cordials was available in the boudoir.

9.419 PLUM (Soft Drink)

A set of three enamel labels for PLUM (fig. 9.205), RASPBERRY and MARMALADE with red borders, a red rose above the title and a yellow rose below it, and with titles in distinctive black lettering, probably refers to fruit juices.

Fig 9.205

9.420 POISON (Medicinal)

Mentioned by Penzer in his Table III at Number 26, POISON probably referred to LAUDANUM. It was engraved on a silver unmarked label. Its eye shape with double reeded border dates it to around 1780 to 1790. It measures 4.2cm. x 2.3cm. and is numbered 144 in the Marshall Collection. See above LAUDANUM and below POISON LAUDANUM.

9.421 POISON LAUDANUM (Medicinal)

This label was hung around a blue bottle in the Medicine Cabinet or travelling chest containing Tincture of Opium. An unmarked silver label of around 1830 engraved POISON LAUDANUM could be one of a set of at least seven perhaps made by Charles Reily and George Storer in the Marshall Collection. POISON LAUDANUM is numbered 347. It is noted in the Journal that this label was "part of a series found in a travelling medicine chest which Mrs Marshall once came upon with all its bottles and labels intact." It could have been an apothecary's chest or a family travelling box. An apothecary would have used hand-written paper labels. So it would seem that it would have formed part of a family's medical travelling box set. This title was incised on a rectangular silver label with rounded corners. All seven labels measure 2.6cm. by 1.4cm. Dr. Buchan writing in 1785 in his "Domestic medicine" gives instructions about the preparation of POISON LAUDANUM : "Take two ounces of crude opium, ten ounces of spirituous water and ten ounces of mountain wine. Dissolve the opium, sliced, in the wine, with a gentle heat, frequently stirring it. Afterwards add spirituous aromatic water and strain. Twenty five drops of the tincture should contain about a grain of opium".

9.422 POMADE (Toiletry)

Originally a scented ointment for application to the skin, POMADE was developed into a lotion for massaging the scalp to stimulate hair growth, prevent baldness, restore the hair and create a glossy effect. Recipes for Orange Pommade, Pommade Divine and Pomatum are given by Dr. Scott in his Toilette No. 1. A formula for Pomade Divine may be found in the Household Cyclopedia of Perfumery. A paper label for Lavender Pomade was affixed to a stoppered blue glass bottle housed in a beautiful silver holder made in Germany around 1890.

9.423 POMAR (Toiletry)

See above, POMADE.

9.424 POMARD (Toiletry)

See above, POMADE.

9.425. POMERANS (Toiletry)

POMERANS, a variety of citrus fruit known as aurantium Amora-Grappen, is used for skin care. The name, along with three variant spellings of it, is recorded in the Master List. POMERANS or POMERANZ should be distinguished from POMMERANS or POMMERANZ which is a BRANDY based liqueur with a bitter taste distilled from fermented fruit as an essence or Schnapps.

9.426 POMERANZ (Toiletry)

See above, POMERANS.

9.427 PONDS EXTRACT (Toiletry)

This title appears engraved upon a plated label once in the Clemenson-Young Collection, probably one of a set of five along with CITRATE MAGNESIA, LYSTERINE, GLYCO-THERMALINE and BORACIC WATER (97). That PONDS EXTRACT was a liquid preparation is shown by a light-blue glass bottle bearing the title in moulded glass.

9.428 PORTUGAL (Perfume and Toiletry)

See above EAU DE PORTUGAL. A silver-gilt label for PORTUGAL, a pair with JESSAMINE, was made around 1820 by John Reily. It measures 2.6cm by 1.5cm and is numbered 576 in the Marshall Collection. An enamel label is illustrated in fig. 9. 206.

Fig 9.206

9.429 POTASH (Medicinal)

Mentioned in the Journal and possibly a pair with SODA, POTASH has powerful caustic properties like caustic SODA. A crude form of potassium carbonate obtained from vegetable ashes, hydroxide of potassium is a hard white brittle substance soluble in water.

9.430 POUSSE L'AMOUR (Perfume)

This was a fragrance. The title appears upon a French enamel, rectangular in shape, with concave top and bottom, having irregular wavy edges, a pink border and floral decoration, which would have looked good on the dressing table. It also, not surprisingly, gave its name to a flavoured cocktail based on cognac, marasquino and crème de cacao, with an added egg yolk (98).

9.431 Pt George (Perfume)

This enamel label is a pair with Limoux. See further above, Limoux.

9.432 PRUNES (Medicinal)

PRUNES was shown on a white oval French enamel with a frilly base and significant floral decoration. Prune juice was taken to avoid or deal with constipation. Its use as a mild laxative has long been recognised. In 1683 an apothecary with the initials TG purchased a jar of Electuarium Ciacassiae cum Manna which was based on prunes with added violet flowers, cassia, sugar of violets, syrup of violets, tamarinds, sugar-candy and manna (99).

9.433 P.VINEGAR (Medicinal, Soft Drink and Toiletry)

A beautiful neck-ring label (see fig. 9.207) by Josiah Snatt dated 1804 with a delicate looking hinge was engraved for P. VINEGAR. It was probably used on a bottle in a soy frame. On the other hand PINK VINEGAR may well have been available in the boudoir. See further below R.VINEGAR and VINEGAR.

Fig 9.207

9.434 QUESTCHE (Soft Drink)
See below QUETSCHE.

9.435 QUETSCH (Soft Drink)
See below QUETSCHE.

9.436 QUETSCHE (Soft Drink)

This was a flavoured cordial. The enamel label is a pair with LAVANDE and is illustrated in fig. 9.208.

Fig 9.208

9.436a QUETSH (Soft Drink)
QUETSH is a variant spelling of QUESTCHE (see paragraphs 9.434 to 9.436 above and paragraph 9.437 below) The label is small and of boudoir size. It bears the name of a plum drink. It is also the name of a spirit where distillation of the fruit has been involved.

9.437 QUUETCHE (Soft Drink)
See above QUETSCHE.

9.438 R.CURRANT (Soft Drink)
This title refers to RED CURRANT which was a fruit juice.

9.439 R.VINEGAR (Soft Drink)
This title refers to RASPBERRY VINEGAR or to RED VINEGAR on both of which see below. The title appears on a plated unmarked octagonal label

with beaded and fluted edges measuring 3.1cm by 2.0cm and numbered 351 in the Marshall Collection.

9.440 RAISEN (Soft Drink)
See below, RAISIN. The title appears on a large plated architectural style rococo label measuring 5.7cm by 4.4cm and numbered 157 in the Marshall Collection.

9.441. RAISIN (Soft Drink)
This title appears on a Battersea enamelled label and then on small silver labels made by Hester Bateman c.1780, Elizabeth Morley in 1804 which was at one time in the Clemson-Young Collection (100), James Atkins in 1805 and John Reily in 1823. A plated RAISIN (fig. 9.209), measuring 3.0cm by 1.9cm, is in set with CHERRY, CURRANT and ORANGE (for each of which please see above). Larger size labels, (such as the two examples illustrated, one with borders of sheaves made by Thomas Wallis and Jonathan Hayne in 1817 (fig. 9.210), the other bearing maker's mark TA and a 1756 lion(fig. 9.210a) and bin labels were used to identify RAISIN WINE which, like CURRANT WINE, was very popular. A bin label can be seen numbered 466 in the Marshall Collection. Spiced up , RAISIN was sold as a substitute for gin, which had been made expensive to buy because of the excise duty of 5 shillings per gallon imposed on gin under the so-called "Gin Act" of 1736. Dr Thomas Wilson, son of the Bishop of Sodor and Man, wrote in his diary for 29th September 1736: "The Gin Shippers now sell a liquor made of Raisin Wine and Currant Wines which they make very strong with ginger and other hot spices but the poor wretches say it makes them sick and not drunk which their beloved gin did. I hope the next generation will drink ale as their forefathers did, who worked as hard and

Fig 9.209

Fig 9.210

Fig 9.210a

lived longer". The Inventory of Matthew Boulton's cellar at Soho House in 1804, when England was at war with France, shows that he had stocked 378 bottles of PORT, 173 of MADEIRA, 121 of LISBON, 78 of SHERRY and 48 of RAISIN, whereas he only had 4 of CLARET, 7 of RUM and 7 of HOLLANDS. Most RAISIN labels are therefore wine labels except perhaps those of small size already mentioned and perhaps also a small plated crescent with its horns embellished by delicate decoration (101) and a small rectangular silver label with cut corners and double reeding made by Thomas and James Phipps in 1820 measuring 4.1cm by 2.3cm and numbered 158 in the Marshall Collection. Two enamel boudoir labels are however illustrated (figs. 9.210b and 9.210c)

Fig 9.210b

Fig 9.210c

9.442   RASBERRY (Soft Drink)

See below, RASPBERRY. The RASBERRY label in the M.V. Brown Collection at the London Museum is large, measuring 5.8cm in width, and was probably made

Fig 9.211

for RASBERRY WINE. So also was a smaller, plated label of oval shape with beaded edge, measuring 4.9cm by 3.9cm and numbered 159 in the Marshall Collection. However, a silver boudoir label with this title was made by Thomas Watson & Co. of Sheffield in 1806. It is of crescent shape simply decorated with a singe reed and measures only 3.6cm by 1.0cm. It is numbered 566 in the Marshall Collection. An enamel label with no grapes is illustrated in fig. 9.211.

9.443   RASIN (Soft Drink)

See above, RAISIN.

9.444   RASPAIL (Medicinal)

Raspail is a tonic in vogue especially between 1850 and 1880. The title appears upon a rather crude enamel. It was yellow in colour and popular with French pharmacists. It was named after Vincent Raspail who first produced it in 1847. The title appears on an enamel label once in Mr J.F. Pidgeon's collection.

9.445   RASPBERRY (Soft Drink)

A set of three enamel labels for PLUM, RASPBERRY (fig 9.212) and MARMALADE with red borders, a red rose above the title and a yellow rose below it, and with

Fig 9.212

distinctive black lettering, probably relates to fruit juices. RASPBERRY might have been a discreet indication of MARASHINO as it was used in its preparation but CHERRY would have been a more appropriate title to use. The title RASPBERRY has been found on an elongated double-sided escutcheon, the reverse having been engraved with DANDELION, attributed to Charles Henry Davis (made in London in 1960), and in set with APRICOT, BLACKBERRY, BLACKCURRANT and GREENGAGE. A small silver RASPBERRY dated 1806 has been noted and the Marshall Collection has a triple reeded plated octagonal label measuring 5.2cm by 2.5cm and numbered 161. Phipps and Robinson made a double reeded rounded ends rectangular silver label with this title in 1797.

9.446   RASPBERRY VINEGAR (Medicinal)

Thought to have been a stimulant. An oval shaped plated label with beaded edge having this title is numbered 162 in the Marshall Collection. It measures 5.5cm by 3.0cm.

9.447   RATAFIA (Medicinal and Soft Drink)

It was quite common to celebrate the legally binding exchange of contracts by drinking a glass of SHERRY or RATAFIA and saying "let it be ratified". Grenoble was famous for its RATAFIAS said to be substantially infusions of fruit. Florence produced its own RATAFIA DE FLORENCE. Menage is of opinion that the term RATAFIA was derived from the East Indies. Leibnitz says that it was a corruption of rectified as in rectified spirits. The Marshall Collection has, numbered 163, a plated label, rectangular in shape with rounded corners and triple reeded borders, which measures 4.5cm by 2.4cm. To-day RATAFIA is the name of a liqueur commercially produced in Danzig. There is also an enamel label for RATAFIA DE FLORENCE (see below). But RATAFIA was also a non-alcoholic cordial like citron water as evidenced by Rone's advertisement for cordials (illustrated in fig. 9.213) sold at his shop near St. George's Church in Southwark. Labelled bottles

Fig 9.213

can be seen in the illustration for USQUEBAUGH, RATAFIA, CITRON WATER and FINE CORDIAL. Part of the apparatus used in the distillation process is also illustrated for USQUEBAUGH (otherwise known in this context as Irish Whiskey), being "green and saffron coloured". Phipps and Robinson made a small silver boudoir label for RATIFIA in 1800, crescent shaped, measuring only 2.4cm by 1.1cm and numbered 409 in the Marshall Collection.

9.448   RATAFIA DE FLORENCE (Soft Drink)
The name of this special kind of RATAFIA (which see above) appears upon an enamel label. This is illustrated in fig. 9.214. Significantly there are no blue grapes forming part of the decoration.

9.449   RATAFIADE FLEUR D'ORANGE (Perfume)
See above, FLEURS D'ORANGE

9.450   RATAFIAT (Perfume)
See above, FLEURS D'ORANGE.

Fig 9.214

Fig 9.215

9.451   Ratafiat de fleur d'orange (Perfume)
This title in script lettering appears upon a French porcelain shield shaped label illustrated in fig. 9.215. See above, FLEURS D'ORANGE

9.452   RATAFIE (Medicinal and Soft Drink)
See above, RATAFIA

9.453   RATIFEE (Medicinal and Soft Drink)
See above, RATAFIA

9.454   RATAFIER (Medicinal and Soft Drink)
See above, RATAFIA.

9.455   RATIFIA (Medicinal and Soft Drink)
In 1800 Phipps and Robinson made a tiny (2.4cm. x 1.1cm.) label for RATIFIA of crescent shape now in the Marshall Collection and numbered 409. See above RATAFIA. A silver label with this title is illustrated in fig. 9.216.

Fig 9.216

9.456   RATIFIE (Medicinal and Soft Drink)
See above, RATAFIA.

9.457   RATTAFIER (Soft Drink)
This title appears on a silver rounded dome crescent label made in London by Crispin Fuller in 1800. See above, RATAFIA.

9.458   RATTFIER (Soft Drink)
See above, RATAFIA.

9.459   R. CONSTANTIA (Medicinal)
This was a boudoir tipple, included in the list because of its medicinal qualities. See below for W. CONSTANTIA and RED CONSTANTIA.

9.460   RED (Writing Ink)
This title is mentioned in the Master List. It may refer to a silver label in set with BLACK or COPYING or some other ink and thus be hung on an ink bottle to show its contents, or it may refer to a RED wine as opposed to a WHITE wine.

9.461   RED CONSTANTIA (Medicinal)
See above, R. CONSTANTIA and see fig. 9.217.

Fig 9.217

9.462   RED CURRANT (Soft Drink)
RED CURRANT juice was a popular drink. See further above for CURRANT.

9.463   RED VINEGAR (Medicinal)
RED VINEGAR was regarded as a useful stimulant. It was often used in the soy frame. The Marshall Collection has a set of unmarked probably English provincial silver labels measuring 2.1cm by 1.1cm with pierced letters for ANCHOVIES, HARVEYS and RED VINEGAR, numbered 331 to 333. However RED VINEGAR has been noted in set with BITTERS. It is a small plated octagonal label with beaded and fluted edges, measuring 3.1cm by 2.0cm. See above BITTERS and below VINEGAR.

9.464   RONDELETIA (Perfume)
Mentioned in Major Gray's list of 1958, a label with the title RONDELETIA is in the Cropper Collection in he V & A Museum. It was a perfume named after a Sixteenth Century French naturalist, one Rondelet, and based on a plant noted by Penzer as being of the tropical American genus of Cinchonaceae. Penzer even discovered a recipe for it given by John Timbs (102) as follows: "Take one gallon of spirit, two ounces of LAVENDER, one ounce of oil of CLOVES, three drams oil of ROSES, one ounce of BERGAMOT, and a quarter of a pint each of Extract of MUSK, VANILLA and ambergris. Mix thoroughly, and in a month it will be fit for use". A similar recipe is given in The Household Cyclopedia of Perfumery (103) for Essence of Rondeletia: "Use spirits (brandy 60 o.p.), 1 gall.; otto of lavender, 2 oz.; otto of cloves, 1 oz; otto of rose 3 drs.; otto of bergamot, 1 oz.; extract of musk; extract of vanilla; extract of ambergris; each a quarter pint. The mixture must be made at least a month before it is fit for sale. Very excellent rondeletia may also be made with whiskey".

9.465   ROSA (Perfume)
See ROSE below.

9.466   ROSE (Perfume)
Rose oil, rose wood oil and rose geranium oil are all natural essential oils. ROSE is a fragrant perfume, popular from the 1860s, made famous by the maceration process used in Grasse in France. Much has been written about the language of flowers, like the language of the lady's fan. Most authors would agree that red roses stand for love and say "I love you", that white roses stand for innocence and purity and say "I am worthy of you", that lavender roses stand for love at first sight and say "I am enchanted by you" and that orange roses stand for fascination and say "I am bewitched by you". ROSE was therefore a very popular perfume of great importance to the perfumier. This aspect is reflected by the illustrated (fig. 9.218a) toilet box perfume silver-gilt label pierced for ROSE. Another toilet box perfume label for ROSE GERANIUM, a pair with EAU DE LAVANDE, is of undoubted quality (104). Titles in the French language followed French leadership in the perfume industry. In a less concentrated form ROSEWATER was very popular. Bottles on the dressing table bore labels in silver for ROSE by Phipps, Robinson and Phipps (fig. 9.218b) in 1812 (similar in design but not a pair with LAVENDER by the same makers and of the same date), and in enamel or porcelain (Crown Staffordshire in the well established design No. A6297) for EAU DE ROSE. See also ATTAR OF ROSES, EAU DE ROSE, MILK OF ROSES, ROSEWATER and WHITE ROSE.

9.467   Rose (Perfume)
See above ROSE. The title Rose appears upon an unmarked silver or silver-plated label illustrated in fig. 9.218c.

9.468   Rosebud Perfume (Perfume)
See above ROSE.

9.469   ROSE GERANIUM (Perfume)
See above ROSE. ROSE GERANIUM is used to-day to flavour a St. Agnes farmhouse ice cream made in the Isles of Scilly.

Fig 9.218a

Fig 9.218c

Fig 9.218b

Fig 9.219a

Fig 9.219b

Fig 9.219c

Fig 9.219d

Fig 9.221

Fig 9.219e

> *To diftil* ROSE WATER.
>
> GATHER your red rofes when they are dry' and full blown, pick off the leaves, and to every peck put one quart of water, then put them into a cold ftill, and make a flow fire under it, the flower you diftil it the better it is, then bottle it, and cork it in two or three days' time, and keep it for ufe.——*N. B.* You may diftil bean-flowers the fame way.

Fig 9.220

9.470 ROSEMARY (Perfume)
ROSEMARY oil is a natural essential oil. ROSEMARY was distilled from the aromatic plant known as ROSEMARY millefoil. This title was engraved on one of the eight silver splayed neck ring labels adorning one of the set of eight glass perfume bottles contained in a shagreen travelling perfume box case. ROSEMARY may have been similar to HUNGARY WATER as they were both prepared from rosemary flowers.

9.471 ROSE WATER (Perfume and Toiletry)
A recipe for the distillation of ROSE WATER is illustrated in fig. 9.220. The title also appears on a boudoir-style label similar to that illustrated in fig. 9.221e for SHERRY. See below ROSEWATER and figs. 9.219a-e and fig. 9.221.

9.472 ROSE.WATER (Perfume and Toiletry)
See below ROSEWATER and above for NOYAU where the circa 1830 tiny label by Reily and Storer is described. ROSE.WATER appears on a plain silver unmarked octagonal label with attractive pierced lettering, measuring 2.9 cm by 1.9cm, hung by a traditional silver chain.

9.473 ROSE-WATER (Perfume and Toiletry)
See below ROSEWATER. The title appears on a two cupids silver label in the Cropper Collection (105).

The title may have been re-engraved and the label re-used to decorate a dressing table scent jar and create interest in the boudoir.

9.474 ROSEWATER (Medicinal, Perfume and Toiletry)

ROSE WATER was distilled. According to an old recipe one should "gather your red roses when they are dry and full blown, pick off the leaves, and to every peck put one quart of water, then put them into a cold still, and make a slow fire under it, the (longer) you distil it the better it is, then bottle it, and cork it in two or three days' time, and keep it for use". It appeared in Major Gray's 1958 list. It refers to an unmarked silver octagonal. Numerous examples of labels can be quoted. By way of examples, ROSEWATER was a label in the Cuthbert Collection; a porcelain label in pair with GLYCERINE is of domed crescent shape with a pink rose on the dome; there are silver-gilt labels of oblong shape, some with beaded edges; a shaped silver label was made by Rawlings and Summers in 1836; various other silver examples are dated 1820, 1832 and 1890. Labels for the concentrated perfume are smaller than those for toiletry. ROSEWATER was used as a light astringent tonic rather like witch hazel. It was also used as an eye lotion because of its mildly astringent properties. It formed part of a German medical box (106) containing 15 silver-mounted bottles labelled and inscribed in German such as Rosen Wasser for ROSEWATER and Wund Wasser for Wound Water or ARQUEBUSADE (on which see above). Gentlemen used ROSEWATER as a toiletry. A perfume bill delivered to King George IV (1820-1830) shows that he spent as much as £500 17s 11d on one occasion on such toiletries including ROSEWATER. Labels were heavily used. ROSEWATER is a badly damaged eighteenth century shaped enamel label with pink borders and with painted yellow and small blue flowers (107). For a long time ROSEWATER has been used as a cleansing agent for washing hands before meals, for example in Livery Company Halls. It was poured into a bowl from a ewer. After a meal the ROSEWATER bowl is circulated. One dips one's napkin into the bowl and applies the ROSEWATER (fig. 9.221) behind the lobes of one's ears. It is supposes to assist digestion. So in 1573 the Master Mercer, William Birdle, gave to his Company a splendid silver wagon which had been made in 1554. This trundled around the dining table carrying a large cask or tun which enabled ROSEWATER to be sprinkled over the hands of diners.

9.475 ROSE WATER & GLYCERINE (Perfume and Toiletry)

This title appears on a pink enamel label. See above ROSEWATER and GLYCERINE (fig. 9.222).

Fig 9.222

9.476 Rp BERRY (Soft Drink)

This title appears on a single reeded octagonal unmarked silver label. See above RASBERRY and RASPBERRY.

Fig 9.223

9.477 SAL VOLATILE (Medicinal and Perfume)

Josiah Snatt made a small travelling box silver label with this title in 1801 (fig. 9.224). An aromatic solution of ammonium carbonate was used as a restorative in fainting fits. Literally meaning (since 1654) salt liable to or susceptible of evaporation and diffusion at ordinary temperatures, it was commonly known as smelling salts. A paper label with this title was included in the Belfast medical box.

 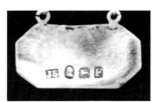

Fig 9.224

9.478 SAVRON (Perfume)

SAVRON is a kind of COLOGNE. The title appears on a French enamel with blue decoration (108). Cosmydor's paper label for Savon a l'Heliotrope from Paris is illustrated.

Fig 9.225

9.479 S.WATER (Medicinal, Perfume and Soft Drink)

Thought to stand for Sweet Water (a charming light perfume), Soda Water (soft drink), Seltzer Water (soft drink) or Surfeit Water (medicinal) this title appears on a silver label of eye shape with an enamel insert made by James Hyde (1774-1799) although it has been attributed to John Holloway (1772- 1795). This title is included in a list published in the Journal (109). S.WATER has been described as containing an essence of musk, cloves, vanillas, benzoin, orange-flower and alcohol. See fig. 9.226.

Fig 9.226

9.480 SCUBAC (Perfume)

This title appears on a late Eighteenth Century enamel label, a pair with EAU DE CELADON, which were possibly comprised in a set of six perfume labels. SCUBAC is also an abbreviated form of Usquebac or Usquebaugh which is Gaelic for Water (Uisge) of Life (Beatha). See further EAU DE CELADON above.

9.481 SCYRUP (Medicinal)

This is thought to be a variant spelling of syrup. A plated label bears the title of SCYRUP in pierced lettering. It is decorated with a border of vine leaves and grapes and is of a large size measuring 6.2cm by 5.2 cm . It is numbered 319 in the Marshall Collection and is a pair with ANNISCETTE, on which see above. SCYRUP could have described a mixture known as Rossolio which was used medicinally to invigorate. One recipe for this household restorative instructs one to take equal quantities of eau-de-vie and Spanish wine and then infuse the mixture with anise, coriander, fennel, citron, angelica and sugar candy dissolved in camomile water. One then boils the mixture until it becomes a thick syrup. See above for GUM SYRUP and MAPLE SYRUP.

9.482 SELTER (Soft Drink)

See SELTZER WATER below. The water originally came from springs in the villages of Selters and Niederselters (near Wiesbaden) in the districts of Hesse-Nassau and Limburg-Weilburg in Germany. It was so popular that various imitations of the original spring water were produced. The title SELTERS appears on an unmarked silver label of inverted scroll, ribbon or banner design.

9.483 SELTERS (Soft Drink)
See SELTER above and SELTZER WATER below. The title SELTERS appears upon an unmarked silver label of inverted scroll, ribbon or banner design. It is early 20th Century.

9.484 SELTZER W. (Soft Drink)
See SELTER above and SELTZER WATER below.

9.485 SELTZER WATER (Soft Drink)
A label with this title is in the Cropper Collection in the V & A Museum (110). SELTZER WATER is an effervescent mineral water originally obtained (or an imitation of it) from a site at Nieder-Selters near Wiesbaden in Germany containing sodium chloride and small quantities of sodium, calcium and magnesium carbonates. In 1758 SELTZER WATER was sold in large bottles for a guinea a bottle by Thomas Davis, purveyor of mineral waters to King George II. See above, SELTER.

9.486 SEVE (Perfume)
This title appears on a double-reeded octagonal silver-gilt unmarked perfume label. The label is small, measuring 2.4cm. x 1.4cm. , and would have been displayed on a small perfume bottle. Victoria Gobin-Daude won a prize for her SEVE exquise in 2002. It was pale green in colour, said to have the milky-sweet smell of tender shoots in the Spring. It is a refreshing toilet water, with reasonable lasting power, giving a feeling of well-being and abundance of life. What more could one ask for?

9.487 SHERBERT (Soft Drink)
See below, SHERBET

9.488 SHERBET (Soft Drink)
This title appears on a silver label made by Joseph Willmore of Birmingham in 1839. SHERBET in

Fig 9.227

Turkish means a drink prepared with Rose Hips. In Arabic it means soda, an effervescent soft drink. The label is die pressed,

similar to others produced by Willmore in 1839. The 1839 date letter is accompanied by a duty mark punch showing the head of King William IV, as the new Queen Victoria punch was still awaited for use. The maker's mark is badly struck. SHERBET was used by nursing mothers to increase lactation (fig. 9.227).

9.489 SODA (Medicinal and Soft Drink)
SODA is sodium bicarbonate (see BICARBONATE OF SODA above and SODA WATER below). It appears in a list published in the Journal (111). It is also mentioned in the Journal (112) as a possible pair with POTASH (which see above).

9.490 SODA WATER (Medicinal and Soft Drink)
SODA WATER is water containing a solution of sodium bicarbonate or charged under pressure with carbon dioxide, strongly effervescent, used as a beverage or stimulant from the 1840s.

9.491 SOURING (Medicinal)
This is a milk product fermented with lactic acid. It was used to improve digestibility. It certainly improved the taste and made food have a longer shelf life. The title appears upon a small cut-cornered rectangle made in Edinburgh around 1800 by William Robertson.

9.492 SPIRIT LAVENDER (Medicinal and Perfume)
This had a medicinal application as can be seen by the nature of the titles of labels in the same set, i.e. CAMPH DROPS, CAMPHtd SPIRIT, EAU D'ALIBURGH, ESSENCE OF GINGER, POISON LAUDANUM and TOOTH MIXTURE. The title appears upon an unmarked silver label with incised lettering, rectangular in shape with rounded ends, double reeded decoration, measuring 2.6cm by 1.4cm and numbered 348 in the Marshall Collection. See above, LAVENDER

9.493 SPRUCE (Medicinal and Toiletry)
This essential oil, made from the leaves of the spruce tree, was used to make a medicinal syrup, not unrelated in substance form to chewing gum and its properties. It was used in the making of SPRUCE beer. It seems to have been paired with LIMOSIN, both being silver plated labels, and was once in the Ryall Collection.

9.494 Star Mise (Perfume)

Star Mise or star-anise is the fruit of the illicium, a Chinese evergreen tree of the magnolia family. It was used both as a perfume and as a flavouring. It imported the smell of aniseed when sniffed or in cooking. The title, being shown on a tiny escutcheon with a small length of chain was probably for a label to be used in a travelling toilet box. Star Anise was used in America to flavour CHERRY Bounce and with ANGELICA in the production of Kummel. See further ANISEED and NOYAU above.

9.495 STRAWBERRY (Soft Drink)

This title for a fruit juice is on an enamel label in the V & A Museum Collection (113). It is probably the engraved title on a glass decanter illustrated by Robin Butler (114) as "S:A:BERRY" when the engraver ran out of space and was forced to abbreviate. The title could perhaps stand for STRAWBERRY Cider which is drunk in the West of England.

9.496 SULTANA (Medicinal and Soft Drink)

This title for a juice has been noted in the Journal as being on an Old Sheffield Plate label (115).

9.497 SURFEIT WATER (Medicinal)

John Farley has given us a recipe for SURFEIT WATER which was used for the relief of upset stomachs. Illustrated are recipes for King Charles II's SURFEIT-WATER and Mr Denzil Onslow's SURFEIT WATER. It was sold in bottled form as a pick-me-up. A glass bottle with this label was sold in the Ashburnham Place sale (116). Mrs. Rhodes when writing in the Journal about a recipe for making it published in Edinburgh in 1759 noted that a CORDIAL of the same name was still being made by a firm in Bristol. According to this recipe to distil cold SURFEIT WATER one takes "two handfuls of spearmint, two of BALM, one of ANGELICA, one of wormwood, one of carduus and one of marigold flowers. Cut them and put them in water; then wring them out and put them in a still; keep wet cloths around it and a slow fire under it." Household home-made recipes were also available (117).

9.498 SWEET (Perfume)

This may stand for Sweet Water which is a charming light perfume. An example is by Robb and Whittet of Edinburgh circa 1860 (see fig. 9.229a).

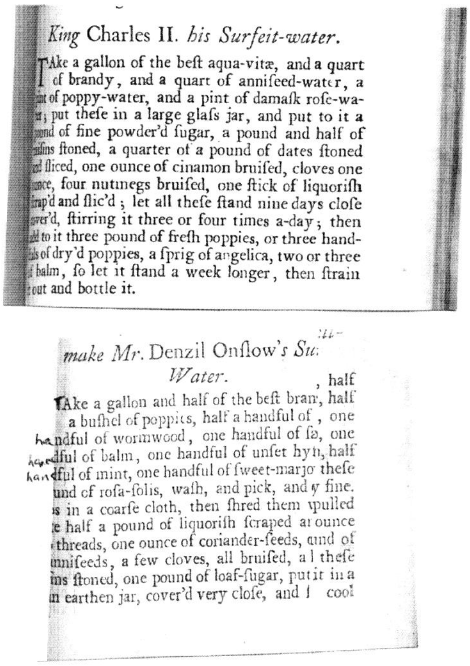

King Charles II. *his Surfeit-water.*

TAke a gallon of the beſt aqua-vitæ, and a quart of brandy, and a quart of anniſeed-water, a pint of poppy-water, and a pint of damaſk roſe-water; put theſe in a large glaſs jar, and put to it a pound of fine powder'd ſugar, a pound and half of raiſins ſtoned, a quarter of a pound of dates ſtoned and ſliced, one ounce of cinamon bruiſed, cloves one ounce, four nutmegs bruiſed, one ſtick of liquoriſh ſcrap'd and ſlic'd; let all theſe ſtand nine days cloſe cover'd, ſtirring it three or four times a-day; then add to it three pound of freſh poppies, or three handfuls of dry'd poppies, a ſprig of angelica, two or three of balm, ſo let it ſtand a week longer, then ſtrain it out and bottle it.

make Mr. Denzil Onſlow's *Su*
Water. , half

TAke a gallon and half of the beſt bran', half a buſhel of poppies, half a handful of , one handful of wormwood, one handful of ſa, one handful of balm, one handful of unſet hyſſ, half handful of mint, one handful of ſweet-marjo theſe und of roſa-ſolis, waſh, and pick, and y fine. s in a coarſe cloth, then ſhred them pulled e half a pound of liquoriſh ſcraped an ounce threads, one ounce of coriander-ſeeds, and of anniſeeds, a few cloves, all bruiſed, a l theſe ins ſtoned, one pound of loaf-ſugar, put it in a n earthen jar, cover'd very cloſe, and l cool

Fig 9.228

9.498a Sweet (Perfume)

See above, SWEET, and fig. 9.229b.

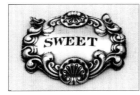

Fig 9.229a

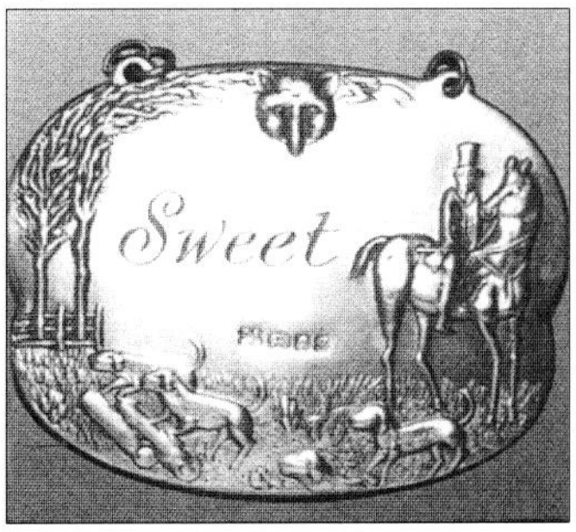

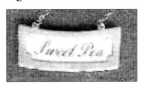

Fig 9.229b

Fig 9.230

9.499 Sweet Pea (Perfume)

This pewter perfume label, one of a set of four, was retailed by J. Floris and Company, purveyors of the finest perfumery and toiletries to the Court of St. James since 1730. The company has a Royal Appointment as creators and suppliers of fragrances.

9.500 TANGERINE (Soft Drink)

The orange-coloured citrus fruit TANGERINE (Citrus tangerina), a variety of the mandarin orange (Citrus reticulata), is well known in the fruit basket. It is a good pick me up drink being a source of vitamins B, B2, B3 and C. Tangerine oil has limonene as its major constituent as do all other citrus oils. Fresh tangerine juice could well have been made available as a breakfast refreshment. The title appears on a silver label made by Thomas Henry Francis and Frederick Francis in 1863 now in the Cropper Collection (118). See fig. 9.231.

Fig 9.231

9.501 TAR WATER (Medicinal)
TAR WATER was used for medicinal purposes.

9.502 THISTLE JUICE (Soft Drink)
This presumably was rather like a herbal drink.

9.503 TISSANNE (Medicinal and Soft Drink)
A silver label with this title, hallmarked in Dublin by an Irish maker, has been noted in the Journal (119). A TISSANNE was used for making a herbal drink by infusion. It was often a kind of pick-me-up. The Ormonde Blyth Collection has a TISANE de CHAMPAGNE. PEPPERMINT tea has been described the finest of fine TISANES by Newby.

Fig 9.232

9.504 TOAST (Medicinal)
This label may have been used to identify prepared medicines ready for taking. There is a reference in the Journal to such medicines being taken by mouth.

9.505 TOILET LOTION (Toiletry)
This title has been shown on a porcelain label, a pair with GLYCERINE (fig. 9.233). It could consist of an aromatic vinegar used as an emollient.

Fig 9.233

9.506 TOILET POWDER (Toiletry)
This title has been noted as appearing on two pink enamel labels (fig. 9.234a and 9.234b).

Fig 9.234a

Fig 9.234b

9.507 TOILET VINEGAR (Toiletry)
This title relates to an aromatic vinegar used as an emollient. It appears upon a plain octagonal silver label made by William Summers in 1885. See also AROMATIC VINEGAR.

Fig 9.235

9.508 TOILET WATER (Toiletry)
Mentioned in Penzer's Table III at No. 29, TOILET WATER is a generic name for perfumes of reduced strength. TOILET WATER appears on an enamel on copper label with pink roses and green foliage on a white background, being part of a set with ALCOHOL, ALMOND CREAM, BORACIC ACID and GARGLE. It also appears on a broad open crescent silver label (fig. 9.236a), with a point 925 silver standard impressed mark, in set with COLORLESS HAIR TONIC, (fig. 9.236b).

Fig 9.236a

Fig 9.236b

9.509 TONIC (Medicinal and Toiletry)
This title appears on a French Samson enamel label. There is however no dot over the letter "I". Nonetheless its size and shape (approximately 4.1cm. x 4.2cm.) are virtually identical to a Samson blank in the V & A Museum. See fig. 9.237.

Fig 9.237

9.510 TOOTH MIX: (Toiletry)
See TOOTH MIXTURE below.

9.511 TOOTH MIX 1829 (Toiletry)
See TOOTH MIXTURE below.

9.512 TOOTHMIX (Toiletry)
The silver label with this title, dated 1829, is numbered 608 in the Marshall Collection. It is said to have been made by Thomas Streetin and is in pair with EAU DE COL. It is crescent shaped with a double reeded border and measures 2.0cm by 1.1cm. See TOOTH MIXTURE below.

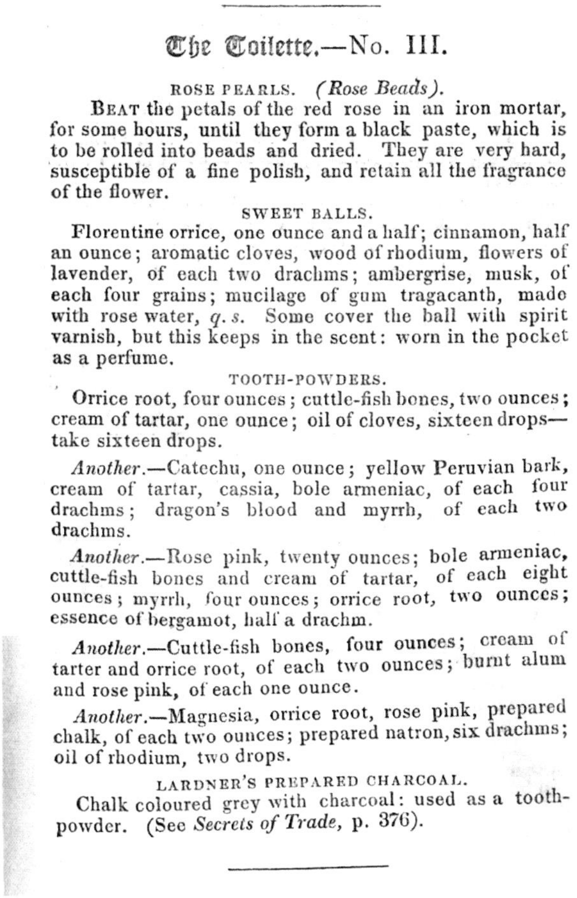

The Toilette.—No. III.

ROSE PEARLS. *(Rose Beads).*

BEAT the petals of the red rose in an iron mortar, for some hours, until they form a black paste, which is to be rolled into beads and dried. They are very hard, susceptible of a fine polish, and retain all the fragrance of the flower.

SWEET BALLS.

Florentine orrice, one ounce and a half; cinnamon, half an ounce; aromatic cloves, wood of rhodium, flowers of lavender, of each two drachms; ambergrise, musk, of each four grains; mucilage of gum tragacanth, made with rose water, *q. s.* Some cover the ball with spirit varnish, but this keeps in the scent: worn in the pocket as a perfume.

TOOTH-POWDERS.

Orrice root, four ounces; cuttle-fish bones, two ounces; cream of tartar, one ounce; oil of cloves, sixteen drops— take sixteen drops.

Another.—Catechu, one ounce; yellow Peruvian bark, cream of tartar, cassia, bole armeniac, of each four drachms; dragon's blood and myrrh, of each two drachms.

Another.—Rose pink, twenty ounces; bole armeniac, cuttle-fish bones and cream of tartar, of each eight ounces; myrrh, four ounces; orrice root, two ounces; essence of bergamot, half a drachm.

Another.—Cuttle-fish bones, four ounces; cream of tarter and orrice root, of each two ounces; burnt alum and rose pink, of each one ounce.

Another.—Magnesia, orrice root, rose pink, prepared chalk, of each two ounces; prepared natron, six drachms; oil of rhodium, two drops.

LARDNER'S PREPARED CHARCOAL.

Chalk coloured grey with charcoal: used as a tooth-powder. (See *Secrets of Trade,* p. 376).

Fig 9.238

Fig 9.239a

9.513   TOOTH MIXTURE (Toiletry)

Mr. Keypstick, described as an Eighteenth Century German quack, sold "teeth powders and cosmetic washes" to ladies. Recipes for various tooth powders were given by Dr. Scott in his "The House Book" in 1826 (see fig. 9.238). Mixtures for the relief of toothache or for cleaning teeth go under a variety of names. CLOVES for example could refer to oil of cloves good for subduing pain. DENTIFRICE was used to clean teeth (see above). Various sorts of GARGLE were used during the grooming process. TOOTH MIXTURE therefore might be almost

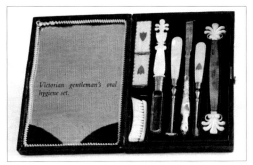

Fig 9.239b

anything from cloves to alcohol. It is likely that it was for relieving toothache rather than for cleaning teeth. Around the 1850s attention to teeth was limited to the more responsible. Silversmiths were involved in making advanced dental instruments. Extraction of teeth saved money on future maintenance. Gentlemen travelled with their own personal hygiene kit. This would include bottles or pots of tooth cleaning powder. Early versions were apparently usually composed of brick dust or charcoal! Examples can be seen in the British Dental Association's Dental museum (figs. 9.239a, 9.239b). TOOTH MIXTURE is incised upon a double reeded rounded end rectangular label measuring 2.6cm by 1.4cm numbered 349 in the Marshall Collection.

9.514   TWININGS BLENDING ROOM 1706 (Soft Drink)

Since Thomas Twining (illustrated in fig. 9.240) opened the first known tearoom (then called a blending room) in 1706, the firm called Twinings has been inextricably linked with British tea-drinking tradition. Twinings' Classics Traditional English tea has been supplemented with, amongst others, Twinings' Aromatics Earl Grey, a light tea which is pale gold in colour and flavoured with the citrus fruit bergamot; Twinings' Moment of Calm flavoured with the citrus zest of orange, the sweet exotic taste of mango and the spiced taste of cinnamon; and Twinings' Fresh and Fruity, having a rose blush colour and flavoured with sweet blackcurrant, ginseng root and vanilla.

Fig 9.240

9.515 VANILLE (Perfume)

The title is French for vanilla, an essential oil. Vanilla is the aromatic substance obtained from the slender pod-like capsule of Vanilla planifolia, a popular perfume since 1728. It appears on a heart-shaped label made out of ivory. This design makes this particular label more likely to be a boudoir label but it should be mentioned that there is a vanilla flavoured liqueur.

9.516 VERBENA (Medicinal, Perfume and Toiletry)

VERBENA, in French known as VERVIENE or VERVEINE, is a toilet water label comprising a combination of oils of lemon peel, orange peel and lemon grass mixed with rectified spirits. The small kidney shaped silver label for VERBENA made by Margaret Binley (1764-1778) was sold as part of a set with EAU DE COLOGNE, AROMATIC VINEGAR and MILLE FLUERS (sic) in 2000 (120). If these small silver labels were made for small bottles housed in a travelling toilet box then they probably related to perfumes. If they were made for slightly larger bottles which were placed on dressing tables then perhaps they were low strength toilet waters or medium strength perfumes used for refilling vinaigrettes. Enamel labels often with titles in French were designed to stay in the boudoir. An example, with the title of VERVEINE, is an enamel label in the Marshall Collection of large size, escutcheon in shape, measuring 5.1cm. by 4.0cm. , has floral decoration suitable for use in the boudoir. An example is illustrated in fig. 9.241 having the characteristic dot over the letter "I" indicating a French provenance or connection. VERVEINE is the title of a French Dreyfous enamel on copper label, slightly the worse for wear, found in Riquwihr, Colmar, Alsace (121). Dreyfous had an English outlet being a shop at 99 Mount Street in the heart of fashionable Mayfair. It has been pointed out (122) that there are three possible uses of VERBENA as (i) a symbolical herb known as VERBENA officinalis with magical powers to resist unfriendly enchantment, (ii) a medicinal potion known to Culpepper as a kind of panacea, for example an aphrodisiac and a cure for migraine, haemorrhoids, dyspepsia, bronchitis, snake-bite and insanity, or (iii) an aromatic known as lippia citriodora or lemon VERBENA which is used in pot-pourri, in sweetening linen, as a scent for soap and most importantly as a toilet water. There are it seems some

Fig 9.241

Fig 9.242

250 different species of VERBENA but only a few are in active cultivation including those of interest to gardeners.

9.517 VERVEINE (Medicinal and Perfume)

VERVEINE has been said to be a finer quality VERBENA, being made with oil of CITRON, FLEUR D'ORANGE, tubereuse, and ESPRIT DE ROSE added. It was retailed as an Eau de Toilette by Messrs. J.Floris in 1905. A beautiful escutcheon shaped enamel with floral decoration bears this title, measuring 5.1cm by 4.0cm, is numbered 429 in the Marshall Collection. Another label, illustrated in the Journal (123), is marked on its reverse DREYFOUS. It is of similar design to a label for ANNISETTE also marked DREYFOUS, being a vase shaped enamel with single roses drooping from the eyelets and a central posy all of which is very feminine. Another VERVEINE in the Gilmour Collection (124) appears to be an unmarked enamelled on copper usual style escutcheon. See above Eau de Toilette and VERBENA and fig. 9.241.

9.518 VERVIENE (Medicinal and Perfume)

This title is included in the Master List. See above VERBENA.

9.519 VICHY (Soft Drink)

VICHY is the name of a town in the Department of Allier in the middle of France which is used from about 1858 to designate a mineral water obtained from springs there.

9.520 VIELLE FINE (Medicinal and Perfume)

A small French enamel escutcheon has this title (fig. 9.242). It is marked "Made in France" on the reverse, perhaps for the English market. See below, VIELLEFINE.

9.521 VIELLEFINE (Medicinal and Perfume)

This one word title appears firstly on a semi-circular French escutcheon decorated with flowers and strokes of blue (hence suitable for the boudoir) in pair with the one word FINENAPOLEON (noted in the Master List Second Update), and secondly on a semi-circular shaped enamel decorated with flowers and measuring 3.5cm by 2.3cm.

9.522 VINAIGRE (Medicinal and Toiletry)

Herbert Lambert made this label in 1908 as a pair with HUILE. Although probably made for an oil and vinegar frame, the use of the French language on silver which was not exported to a French speaking country suggests that the labels could be connected with a travelling box and this concept is supported by the choice of silver-gilt.

9.523 Vinaigre Ordinaire (Medicinal and Toiletry)

A French porcelain shield-shaped label, a pair with Graine-de-lin (which see above) bears the script title Vinaigre Ordinaire. See further above AROMATIC VINEGAR and TOILET VINEGAR.

9.524 VINEGAR (Medicinal and Toiletry)

Small VINEGAR labels were usually made for and used on a vinegar bottle in an oil and vinegar frame or in a soy frame. Large VINEGAR labels were made to identify the contents of storage containers or for use in the cellar as in the case of a pottery bin label with rounded shoulders which has the title VINEGAR. Balsamic, Commandaria and Sherry vinegars were aged using the solera process. A third class of VINEGAR labels were made for use in the boudoir. AROMATIC VINEGAR and TOILET VINEGAR are certainly boudoir labels. Plain VINEGAR labels could be boudoir labels or sauce labels. ELDER, FRENCH, ORDINARY, PINK, RASPBERRY, RED and WHITE VINEGARS have

Fig 9.247a

Fig 9.246   Fig 9.247b

Fig 9.247c   Fig 9.247d

been included in the list of boudoir labels so as not to be overlooked. Chili, Chilli and Chilly Vinegar and CHV were probably only used in the soy context and so have not been included in the list. One test to apply is to ascertain whether a vinegar is in set with other well known sauce titles. So in one instance the script French Vinegar is in set with the script Chilli Vinegar and thus almost certainly is not a boudoir label. Another test is size and appearance. Eight VINEGAR labels are illustrated. Some are larger than others and more decorative. The two labels with double reeded borders are more likely to be sauce labels as this was a common design for sauce labels. The tiny plain VINEGAR almost certainly comes from a travelling box and the single reeded but otherwise plain VINEGAR probably also was used on a bottle in a travelling box.

Home-made VINEGARS were best. The 1754 Edition of Stow's Survey refers to the distilling of "Aqua Vitae, Aqua Composita, Vinegar and the like" made out of hogwash and the dregs of wine See further above AROMATIC VINEGAR, ELDER VINEGAR, French Vinegar, Vinaigre Ordinaire, P.VINEGAR, Rp BERRY VINEGAR, RED VINEGAR, TOILET VINEGAR and below WHITE VINEGAR. Illustrated are a silver-gilt wide-open crescent with worn marks (fig. 9.243), a medium opening unmarked crescent (fig. 9.244), a single

Fig 9.243   Fig 9.244

Fig 9.245   Fig 9.245a

reeded octagonal by Joseph Willmore (fig. 9.245) in 1808 (fig. 9.245a), a plain octagonal by George Fisher of London made in 1877 (fig. 9.246), an eye-shaped with wrigglework borders by Thomas Hyde (fig. 9.247a), a gadrooned oval by Mary and Charles Reily (fig. 9.247b), a double-reeded rounded-end rectangular by Charles Rawlings in 1819 (fig. 9.247c) and a double-reeded octagon apparently unmarked (fig. 9.247d).

9.525   VIOLET (Perfume)

VIOLET is a fragrant perfume made famous by the maceration process in Grasse in France. One recipe for VIOLET includes extracts of cassia, ROSE and tuberose mixed

Fig 9.248

with tincture of arris. It is one of a set of three with JONQUIL and JASMIN in the Cropper Collection in the V & A Museum. VIOLET is quite a strong perfume. It pervades. One can still smell it in an empty stoneware Edwardian perfume bottle with a cork stopper decorated with violets, labelled Desert Violets and bearing an impressed mark for "England" (illustrated). It is also in set (fig. 9.248) with CRÈME DE VANILLE, WHITE LILAC and WHITE ROSE measuring 57mm by 39mm approximately . See also below, VIOLETTE DE PARME. An escutcheon shaped paper label for Wood Violet has been noted on a large plain glass bottle.

9.526   VIOLET DE PARME (Perfume)
See below, VIOLETTE DE PARME.

9.527   VIOLETE (Perfume)
See above, VIOLET.

9.528   VIOLETTE (Perfume)
This title was displayed on a tiny single reeded oval silver label (illustrated) made by William Stephen Ferguson in Elgin around 1830 (125). He had moved to Elgin from Edinburgh in 1828. See above, VIOLET and below fig. 9.250.

Fig 9.249

Fig 9.250

9.529   VIOLETTE DE PARME (Perfume)
VIOLETTE DE PARME is a silver label made in Birmingham in 1900, a pair with WHITE ROSE. They are probably toilet box labels. There are thirteen recorded varieties of this perfume, including Ash Vale Blue, Comte de Brazza, Duchesse de Parme, Gloire de Verdun, Hopley's White, Lady Hume Campbell and Nadaline. These perfumes are still made by Nathalie Casbas in Villaudric. Violette Madame was a perfume made by Jacques Guerlain in Paris from 1901. Liberty's reproduced a label for Violette de Nice.

9.530   VIOLLETE (Perfume)
This name has been recorded without any particulars given. There are over 60 named varieties of VIOLETTE in the Nathalie Casbos flower collection. See above, VIOLET.

9.531   WATER (Soft Drink)
This title was in Major Gray's list of 1958. It may possibly have been part of a set with IODINE. It also appears on what would appear to be a silver vine leaf label made by Rawlings and Summers for export to France circa 1844 with maker's mark only defaced by French import marks perhaps for use by an English diplomatic family. The attribution is based upon similar silver-gilt labels pierced for SHERRY (illustrated in fig. 9.252), BUCELLAS, MADEIRA and CLARET, where the import mark has been struck alongside the maker's mark instead of on top of it (126). WATER also appears on a mid-Victorian silver vine leaf which has been attributed to John Dudley of Great Turner Street, London. See further above BARLEY WATER, BORACIC WATER, BRISTOL WATER, DENTRIFICE WATER, DISTILLED WATER, DRINKING WATER, EAU, MALVERN WATER, SELZER WATER, SODA WATER, SWEET Water, SURFEIT WATER, TAR WATER and VICHY.

Fig 9.251

Fig 9.252

9.532 WATER SOFTNER (Soft Drink)
The name has been recorded with no further details being made available. The label would have been used to identify a glass decanter containing a chemical used to deal with the problem of hard water.

9.533 W. CONSTANTIA (Medicinal)
W. CONSTANTIA (White Constantia) is part of a set of five labels made by Margaret Binley around 1770, which included R. CONSTANTIA (Red Constantia), ALE, CYDER and PERRY, which were all boudoir tipples. W. CONSTANTIA is included in the list because of its medicinal qualities. So the silver labels are small narrow rectangulars with rounded ends and feathered borders by way of decoration, and numbered 226 in the Marshall Collection. See also above, R. CONSTANTIA.

9.533a WHITE CONSTANTIA (Medicinal)

A silver eye-shape label is illustrated. See above, W.CONSTANTIA.

Fig 9.253

9.534 W. CURRANT (Soft Drink)
Thought to stand for WHITE CURRANT, which see below.

9.535 WHITE CURRANT (Soft Drink)
White currants were used to make a fruit juice. They were also used to make a home-made wine. The Caledonian Horticultural Society awarded a prize in 1815, being a silver wine label for MADEIRA made by Joseph Angell, to a Mrs. Thomson "for best WHITE CURRANT wine", which is numbered 296 in the Marshall Collection. One clearly had to make do and mend during the wars with France which restricted the import of French wines. See above BLACK CURRANT and CURRANT.

9.536 WHITE LILAC (Perfume)
A label with this title is illustrated in "Wine Labels" at Figure 1013. It is a small delicate enamel label. It is similar in style to EAU DE ROSE (which see above). It was in Major Gray's list. There is a distinctive dot over each of the capital letter "I"s. Of similar style and size is the label shown in "Wine Labels" for METHYLATED SPIRITS. One can conclude that the

Fig 9.254

Fig 9.255

Fig 9.256

WHITE LILAC (fig. 9.255) measuring 5.7cm. by 3.9cm. is a Samson enamel on grounds of design and appearance in set with WHITE ROSE (fig. 9.259), VIOLET and CRÈME DE VANILLE.

9.537 White Lilac (Perfume)
This was one of a set of four pewter labels for perfumes (fig. 9.256). See above SWEET PEA and WHITE LILAC.

9.538 WHITE ROSE (Perfume)
WHITE ROSE was a very popular perfume and several examples of labels can be mentioned. The title appears on an oval shaped porcelain label with a floral border (fig. 9.258). Penzer was aware of an enamel label. He illustrated it on Plate I of his "Book of the Wine Label". There is an enamel example in the V & A Museum. There are Samson examples both small and large sizes. The small size, 3.0cm. by 2.6cm, is a pair with DACTILLUS. These would be suitable adornments for perfume bottles in a travelling case. Although there is no dot over the letter "I", their design and appearance suggest Samson. The large size, approximately 5.7cm. by 3.9cm., was used for the set of four WHITE ROSE (fig. 9.259), WHITE LILAC, VIOLET and CRÈME DE VANILLE, and for a WHITE ROSE with slightly different decoration (fig. 9.257) measuring 5.73cm. by 3.85cm. A silver label for WHITE ROSE was made as a pair with VIOLETTE DE PALME in Birmingham in 1900, probably for a travelling box. The Good Scents Company of Wisconsin has published a material safety sheet for WHITE ROSE perfume oil under Title III of the Superfund Amendment and Reauthorisation Act of 1986. The hazardous nature of the ingredient concerned remains a trade secret.

Fig 9.259

Fig 9.257

Fig 9.258

9.539 WHITE VINEGAR (Medicinal and Toiletry)

This small label may well have been used on a bottle in a soy frame or oil and vinegar frame, but it could have been used in a travelling box. See above VINEGAR.

9.540 WOODBINE (Perfume)

WOODBINE was part of the set of eight silver splayed hoop neck-ring labels made around 1800 which adorned eight perfume bottles located in a shagreen or shark's skin travelling box.

9.541 YLANG-YLANG (Medicinal, Perfume and Toiletry)

Otherwise known as Cananga Odorata, this essential oil derived from the flowers of the fragrant Canaga tree is often used in aromatherapy. One variety is an important source of perfume. This is sometimes called the Macassar oil plant or perfume tree which comes from Indonesia. The flowers are highly scented, drooping, long stalked with six narrow greenish-yellow petals. It is believed to relieve high blood pressure. It was used for travelling as being a motion sickness remedy. It is known for its ability to slow down rapid breathing and is beneficial in dealing with shock, anxiety or anger (127). Seeds from the sapindaceous tree schleichera trijuga were used in the production of a hair oil which led to furniture coverings being protected by the use of anti-Macassars. The name YLANG-YLANG is derived from Taglog, having the meaning of wild and rare.

NOTES

(1)   City of London Museum Collection 4.102.

(2)   7WLCJ8, p184, dealing with star labels.

(3)   Doctor W. Buchan, Domestic Medicine, 1785.

(4)   See further below, AROMATIC VINEGAR.

(5)   See further below, VINEGAR.

(6)   Acqua di Parma Website and Wikipedia.

(7)   R.B.C. Ryall, 2WLCJ3, p42.

(8)   For definition of Master List see under "References".

(9)   See note (8) above.

(10)   1WLCJ4, p48.

(11)   See note (8) above.

(12)   The collection assembled by Mrs Marshall is one of the most important sources of information about boudoir labels. It is housed in the Ashmolean Museum in Oxford. For details see under References on page VIII.

(13)   See note (8) above.

(14)   See note (12) above.

(15)   See note (12) above. See further reports in 2WLCJ9 at pp155 to 158, where the hyphen is used. The original can be seen in the centre of the bottom line of the Frame of 18th Century labels hung on the wall of Room 55 in the Museum. A key plan to this frame is published in the Marshall Collection to assist in judging comparative sizes of displayed labels.

(16)   Shorter Oxford English Dictionary definition.

(17)   Website explanation.

(18)   Bonham's Auction, 20.5.2006, lot 97.

(19)   1WLCJ6, p81.

(20)   Woolley & Wallis Auction, 25.7.2007, lot 513.

(21)   11WLCJ1, p27.

(22)   See Letters of Gui-patin, Vol.ii, p.425.

(23)   See list of thirteen bottles in R.G. Bignall's "A Travelling Cordial Case", 1WLCJ5,p.61.

(24)   4WLCJ7, p132.

(25)   See note (23) above.

(26)   See note (8) above.

(27)   See 10WLCJ10, p303; Christies South Kensington auction, 12.9.2000, lot 45.

(28)   Marshall Collection numbers 222 to 226.

(29)   See illustration of Campaign Chest, with medicine bottle hung with a label marked by William Elliott. Lainé, a well-known gunmaker of Paris, described himself as an "arquebusier".

(30)   See 12WLCJ7, pp 328 and 329.

(31)   See 2WLCJ4, p.63 for Major Gray's List.

(32)   See 7WLCJ10, p229.

(33)   Sketch is taken from 1WLCJ2, p.26. It was included in Mrs. Rhodes' collection and is illustrated in Row A on p.109 of 1WLCJ8.

(34)   See 1WLCJ5, p.69.

(35)   See 1WLCJ15, p.208.

(36)   See 11WLCJ7, p.276.

(37)   See 1WLCJ4, p.45.

(38) Extracted from "Fit for Life II – Living Health" by H and M Diamond.

(39) Major Gray's list of April 1958 published in 2WLCJ4, p63.

(40) Lot 655 reported in 6WLCJ9, p251.

(41) Table III, number 13.

(42) See 12WLCJ9, p503.

(43) 6WLCJ10, p267.

(44) Illustrated in 2WLCJ8, p13, plate 2, row D7.

(45) V & A Museum, M 1561.

(46) V & A Museum, M 1508.

(47) See further the interesting review by Bruce Jones in 5WLCJ3,pp 57-59.

(48) Christies, 12.6.1980, lot 638; 6WLCJ9, p251.

(49) Illustrated in 2WLCJ8, p13, plate 2, row B3.

(50) Woolley & Wallis auction, 18.10.2006, lot 300.

(51) Shown on "Flog It!" transmitted on Sunday 10.4.2011.

(52) Christies South Kensington auction, 13.1.2011, part lot 61.

(53) Christies South Kensington auction, 13.1.2011, part lot 61.

(54) Illustrated in 10WLCJ2, p36, row A.

(55) V & A Museum, M 1451.

(56) Hoffbrand, p38.

(57) Butler, Great British Wine Accessories, p179.

(58) Butler, Great British Wine Accessories, p 29.

(59) Bonham's auction, 4.12. 2007, lot 479.

(60) Illustrated in 2WLCJ8, p131, row B5.

(61) Illustrated by Butler in Great British Wine Accessories at pp 29 and 248 (see top shelf).

(62) Illustrated in 2WLCJ8, p129, row J3.

(63) Information taken from packet of Green tea.

(64) See 11WLCJ1, p33.

(65) Illustrated in 10WLCJ7, p208, row D2.

(66) "Cottage Physician", p465: Take 1.5 ounces of Palma Christi oil mixed with .5 drachm of Oil of lavender and Apply morning and evening for 3 to 4 months to those Places where the hair is wanting.

(67) 9WLCJ4, p97.

(68) 2WLCJ8, p130, row J3, on plate 2.

(69) 7WLCJ4, p103; Sotheby's 15.2.1979.

(70) 1WLCJ8, p115-116, H.E. Rhodes on "Medicinal Labels".

(71) Rosemary Gladstar, Storey Books, 1999.

(72) 10WLCJ4, p93.

(73) See "A Bachelor's Cupboard" by John W. Luce.

(74) Sold by Christies on 4.4.1870.

(75) Woolley & Wallis, auction 30.10.2008, lot 908.

(76) Bonham's auction, 5.10.2010, lot 28.

(77) 1WLCJ2, p23.

(78) Christies auction, 6.10.1983, lot 760.

(79) 1WLCJ1, p8.

(80) 1WLCJ7, p93, row 2.

(81) V & A Museum, M10-1944.

(82) 3WLCJ2, p39.

(82a) See "Limeys", David Harvie, Sun Publishing, 2002

(83) 11WLCJ5, p167.

(84) 2WLCJ8, p130.

(85) V & A Museum, M118-1944.

(86) 10WLCJ8, "A Taste of Punch".

(87) 1WLCJ14, p201.

(88) Woolley & Wallis, auction 29.10.2008, lot 923.

(89) 7WLCJ p131, 10WLCJ p303, Christies South Kensington, auction 12.9.2000, lot 45.

(90) fragrantica.com for MON PERFUM.

(91) Woolley & Wallis, auction 29.7.2009, lot 467.

(92) See further 1WLCJ9, pp125-126.

(93) 1WLCJ7, p97.

(94) Ingoldsby Legends, R.H. Barham 1837, ed. Bentley 1840.

(95) 8WLCJ2, p42.

(96) 9WLCJ8, p210.

(97) 1WLCJ9, p129.

(98) 12 WLCJ7, p371.

(99) The Hoffbrand Jars Collection, Apothecaries' Co.

(100) 1WLCJ9, p128.

(101) Woolley & Wallis, auction 29.10.2008, lot 915.

(102) Penzer, The Book of the Wine Label, p274.

(103) See http://www.mspong.org/cyclopedia/ perfumery.

(104) 5WLCJ2, p46.

(105) V & A Museum, M 925.

(106) Christies Geneva, May 1996. Illustrated in Cummins.

(107) Bonham's auction, 4.12.2007, lot 496.

(108) Shown at WLC's Spring meeting, 2008.

(109) See further 1WLCJ5, p61.

(110) 4WLCJ, p133.

(111) 1WLCJ, p62.

(112) 1WLCJ, p35.

(113) V & A Museum, M 1510.

(114) Butler, Great British Wine Accessories, p129.

(115) 6WLCJ10, p273.

(116) 1WLCJ8, p116.

(117) 8WLCJ1, p16.

(118) V & A Museum, M 1014.

(119) 6WLCJ9, p251.

(120) Christies South Kensington, 12.9.2000, lot 45.

(121) 10WLCJ3, p89.

(122) The detailed reasons are given in Ian Smart's paper of 9.11.1997 reproduced in 10WLCJ, p94.

(123) 10WLCJ3, p90.

(124) Illustrated in 10WLCJ4, at p93.

(125) Bonham's auction, 25.11.2004, lot 505.

(126) Woolley & Wallis auction 29.4.2009, lot 465 and Bonham's auction 5.10.2010, lot 207.

(127) See further 1WLCJ7, p96.

Fig 10.1

CHAPTER 10

SILVER LABELS

10.1 The dates given in this chapter are for the purposes of general guidance. They are not intended to be definitive. The quality of silversmithing is indicated by the unmarked LEMON shown in fig. 10.1.

10.2 SILVER-GILT
Silver-gilt labels relate to perfumes. They convey a sense of well-being. Silver-gilt looks good on exotic perfume bottles. Illustrated are two modern Guerlain bottles with paper labels for AQUA ALLEGORIA and LILIA BELLA (fig. 5.9). Because of the small sizes of perfume bottles which have to fit into travelling cases, the silver labels were small and lightweight. So often they were impressed with a maker's mark only or no mark at all, in which event identification depends upon an appreciation of the design. Armand Gross, for example, produced and marked a slot-label with a design similar to the unmarked BERGAMOT and FRANGIPANI in the Victoria and Albert Museum. It is helpful to have a list of known makers and dates for establishing the historical background. The dates given refer to the approximate working life of the silversmith to establish a dating bracket. Unattributed silver-gilt labels include titles for CAMPHOR JULEP, EAU DE COLOGNE, NOYAU, ROSE, ROSE WATER (two examples) and SEVE. No gold boudoir labels have been noted.

10.3 ALPHABETICAL INDEX OF SOME SILVER-GILT MAKERS

	NAME	DATES	LOCATION		NAME	DATES	LOCATION
1.	GARRARD, ROBERT II	1818-1860	LONDON		LAVENDER WATER	1831	
	NOYAU	1846			LAVENDER WATER	1836	
					FLEUR D'ORANGE	1837	
2.	GROSS, ARMAND	1893-1899	PARIS		PARFAIT AMOUR	1837	
	BERGAMOT	c1895					

Note. Maker's mark on EAU DE ROSE is obliterated. Attribution is based on the choice of Cherub's Head design.

	NAME	DATES	LOCATION
	FRANGIPANI	c1895	

Note. Attributed to this maker on stylistic grounds, a similar design bearing his Mark.

	NAME	DATES	LOCATION
7.	REILY, JOHN	1800-1826	LONDON
	JESSAMINE	c1820	
3.	LAMBERT, HERBERT CHARLES 1902-1916 LONDON	PORTUGAL	c1820

	NAME	DATES	LOCATION		NAME	DATES	LOCATION
3.	LAMBERT, HERBERT CHARLES	1902-1916	LONDON		PORTUGAL	c1820	
	HUILE	c1910			LAVENDER	c1825	
	VINAIGRE	c1910			EAU DE COLOGNE	c1825	
					ESSENCE OF ROSE	c1825	
4.	PHIPPS, THOMAS and JAMES II	1816-1823	LONDON				
	ESPRIT DE ROSE	1820		8.	STORR, PAUL	1800-1838	LONDON
					KIRSCHENWASSER	c1835	
5.	RAWLINGS, CHARLES	1817-1829	LONDON		NOYAU	1835	
	EAU DE COLOGNE	c1825					
	HUNGARY WATER	c1826		9.	WAKELEY and WHEELER	1930-1960	BIRMINGHAM
					CRAB APPLE	1956	
6.	RAWLINGS, CHARLES and	1829-1897	LONDON		ELDERBERRY	1956	
	SUMMERS, WILLIAM				FONTINIAC	1956	
	EAU DE ROSE	1830			MEAD	1956	
	ELDR.FLOR.WATER	1830			PARSLEY	1956	

SILVER LABELS

10.4 ALPHABETICAL INDEX OF SOME SILVER MAKERS

	NAME	DATES	LOCATION		NAME	DATES	LOCATION
1.	ANGELL, JOSEPH	1811-1848	LONDON	11.	BATEMAN, WILLIAM	1815-1827	LONDON
	LEMON	1829			BALM	1822	
2.	ARCHER, HENRY	1910-1920	SHEFFIELD	11a.	BATSON, JOHN & SONS	1880-1892	
	ORANGE BITTERS	1916			ODC	1880	
	ORANGE CURACAO	1916			ODV	1880	
3.	ARMY AND NAVY CO-OPERATIVE SOCIETY	1903-1929	LONDON	12.	BEEBE, JAMES	1811-1837	LONDON
	BRANDY	1914		13.	BENT and TAGG	1832-1850	BIRMINGHAM
					DENTIFRICE WATER	1834	
4.	ASH, JOSEPH	1799-1818	LONDON				
	BRANDY	1799		14.	BINLEY, RICHARD	1745-1764	LONDON
	RAISIN	1805			ELDER	c1760	
5.	ATKINS, JAMES	1792-1815	LONDON	15.	BINLEY, MARGARET	1764-1778	LONDON
	RAISIN	1805			ALBA FLORA	c1770	
					ALBA-FLORA	c1770	
6.	BARKER, SUSANNA	1778-1793	LONDON		AROMATIC VINEGAR	c1770	
	COLARES	c1780			EAU DE COLOGNE	c1770	
	PARSNIP	c1785			ELDER	c1770	
	GINGER ESSENCE	1791			MILLE FLUERS	c1770	
	ESSENCE	1791			R.CONSTANTIA	c1770	
	FRENCH	1793			VERBENA	c1770	
					W.CONSTANTIA	c1770	
7.	BATEMAN, HESTER	1761-1790	LONDON				
	APPLE	c1770		16.	BRASTED, FREDERICK	1862-1888	LONDON
	ORANGE	c1775			MILK PUNCH	1875	
	RAISIN	c1780					
				17.	BRIDGE, JOHN	1777-1834	LONDON
8.	BATEMAN, PETER and ANN	1791-1799	LONDON		BRANDY	c1830	
	CURRANT	1792					
	BRANDY	1798		18.	BRITTON, GOULD & Co	1898-1910	LONDON
					EAU DE TOILETTE VERVEINE	1905	
9.	BATEMAN, PETER, ANN and WILLIAM	1800-1805	LONDON				
	CHERRY	1799		19.	CRICHTON BROTHERS	1895-1956	LONDON
					CHERRY	1953	
10.	BATEMAN, PETER and WILLIAM	1805-1815	LONDON		GINGER	1953	
	METHYLATED	1807		20.	CHRISTENSEN, NICOLAI	1820-1832	COPENHAGEN
					GUJAVA	c1830	

SILVER LABELS

NAME	DATES	LOCATION	NAME	DATES	LOCATION
21. COCKS & BETTRIDGE	1773-1820	BIRMINGHAM	33. EVANS, WILLIAM WALKER	1872-1880	LONDON
CURRANT	1814		BITTERS	1876	
GOOSEBERRY	1814				
			34. FERGUSON, WILLIAM STEPHEN	1828-1875	ELGIN
22. COLLINS, JAMES	1816-1838	BIRMINGHAM	EAU.DE.PORTUGAL	c1830	
PEPPERMINT	1838		VIOLETTE	c1830	
23. CONSTABLE, WILLIAM	1806-1820	DUNDEE	35. FERRIS, RICHARD	1784-1812	EXETER
CURRANT	1807		FRUIT	1796	
			GINGER	1796	
24. CUNNINGHAM, WILLIAM	1807-1828	EDINBURGH			
CURRANT	1807		36. FISHER, GEORGE	1866-1880	LONDON
			VINEGAR	1877	
25. DALGLEISH, CHARLES	1816-1825	EDINBURGH			
FRONTIGNAN	1817		37. FOUNTAIN, WILLIAM	1794-1823	LONDON
			Hartshorn	1822	
26. DAVIS, CHARLES HENRY	1950-1970	LONDON			
DANDELION	1960		38. FRANCIS, THOMAS HENRY &		
RASPBERRY	1960		FRANCIS, FREDERICK	1854-1866	LONDON
			TANGERINE	1863	
27. DILLER, THOMAS	1828-1851	LONDON			
HAIR WATER	1832		39. FULLER, CRISPIN	1792-1823	LONDON
EAU DE COLOGNE	1837		RATTAFIER	1800	
28. DOUGLAS, JOHN	1804-1821	LONDON	40. GIBSON, JOSEPH	1784-1820	CORK
AURQUEBUSADE	1809		C;BRANDY	c1790	
29. DUDLEY, JOHN	1846-1856	LONDON	41. GLENNY, JOSEPH	1792-1821	LONDON
WATER	c1850		GINGER	c1819	
29a. EDWARDS, THOMAS	1816-1836		42. GODBEHERE, SAMUEL, WIGAN,	1800-1818	LONDON
PEPPERMINT	1825		EDWARD and BULT, JAMES		
FONTANIAC	1828		CURRANT	1804	
30. ELLERBY, WILLIAM	1802-1815	LONDON	43. GRAY, ROBERT	1776-1806	GLASGOW
LEMON	1810		CORDIAL	c1805	
31. ELLIOTT, WILLIAM	1795-1809	LONDON	44. GROSS, ARMAND	1893-1899	PARIS
	1813-1830		Silver gilt labels		
ARQUEBUSADE	1818				
			45. HALFORD, THOMAS	1807-1832	LONDON
32. EMES, JOHN	1796-1808	LONDON	HUILLE DE ROYAUX	c1807	
NECTARINE	1799				

SILVER LABELS

NAME	DATES	LOCATION	NAME	DATES	LOCATION
46. HAMILTON & INCHES	1866-2012	EDINBURGH	59. KNIGHT, GEORGE	1818-1825	LONDON
CHILLI VINEGAR	1912		ORANGE	1820	
FRENCH VINEGAR	1912				
			60. KNIGHT, SAMUEL	1810-1827	LONDON
47. HAYNE & Co.	1836-1848	LONDON	GRAPE	1816	
CAMPHORATED SPIRITS OF WINE	1843				
			61. KNIGHT, WILLIAM	1810-1846	LONDON
48. HAZLEWOOD, THOMAS	1888-1915	BIRMINGHAM	NOYEAU	c1830	
METHYLATED	1913				
			62. LAMBERT, HERBERT CHARLES	1902-1912	LONDON
49. HOCKLEY, DANIEL	1810-1819	LONDON	HUILE	1908	
CAPILARE	1818		VINAIGRE	1908	
			Silver gilt labels		
50. HOLLOWAY, JOHN	1772-1795	LONDON			
S. WATER	1788		63. LEA AND CO.	1811-1824	BIRMINGHAM
			COWSLIP	1821	
51. HOUGHAM, SOLOMON	1793-1817	LONDON	CURRANT	c1821	
ARQUEBUZADE	1808				
CURRANT	1812		64. LEVI AND SALAMAN	1870-1921	BIRMINGHAM
			Eyes	1901	
52. HOWDEN, WILLIAM	1801-1828	EDINBURGH			
GINGER	c1805		65. LINTVELD, HERMANUS	1817-1833	AMSTERDAM
			ANISETTE	1825	
53. HUKIN & HEATH	1881-1953	LONDON/	CURACAU	1825	
		BIRMINGHAM			
			66. LINWOOD, MATTHEW II	1793-1821	BIRMINGHAM
ASTRINGENT	1908		ARBAFLOR	1812	
CRÈME DE NOYEAU	1908				
			67. LIVINGSTONE, EDWARD	1792-1810	DUNDEE
54. HYDE, THOMAS II	1747-1804	LONDON	ACID	1795	
VINEGAR	c1775				
			68. LOWE, GEORGE	1791-1841	CHESTER
55. HYDE, JAMES	1774-1799	LONDON	EAU DE COLOGNE	1838	
S. WATER	c1790				
LEMON	1796		69. MANSON, DAVID	1809-1820	DUNDEE
			GINGER	c1810	
56. JOHNSTON, ALEXANDER	1760-1799	LONDON			
BRANDY	1779		70. MAPPIN & WEBB	1859 to date	LONDON
			METHYLATED SPIRIT	1914	
57. KAY, CHARLES	1815-1827	LONDON			
GRAPEFRUIT	c1820		71. MATTHEWS, HENRY	1874-1897	LONDON
			ANNISCETTE	1880	
58. KEHOE, DARBY	1771-1780	DUBLIN	CHARTREUSE	1880	
FRONTINIAC	1775		COGNAC	1880	

NAME	DATES	LOCATION	NAME	DATES	LOCATION
72. MILLS, NATHANIEL	1825-1848	BIRMINGHAM	GOOSEBERRY	1815	
BRANDY	1845		GRAPE	1816	
73. MOODY, WILLIAM	1751-1786	LONDON	81. PHIPPS, THOMAS and JAMES II	1816-1823	LONDON
CURRANT	1785		CURRANT	1816	
			GRAPE	1816	
74. MORLEY, ELIZABETH	1794-1814	LONDON	CURRANT	1817	
LEMON	c1795		ESPRIT DE ROSE	1820	
OIL	c1800		NOYEAU	1820	
RAISIN	1804		RAISIN	1820	
COWSLIP	c1805				
RAISIN	1808		82. PRICE, JOSEPH	1821-1838	LONDON
CURRANTS	1811		Crème de Noyeau	1832	
DAMSON	1811		Crème de The	1832	
GOOSEBERRY	1811				
ORANGE	1814		83. RAWLINGS, CHARLES	1817-1829	LONDON
			VINEGAR	1819	
75. NEWLANDS, JAMES and	1811-1816	EDINBURGH	ELDER FLOWER	c1820	
GRIERSON, PHILIP			OIL	1820	
OIL	1813		ORANGE	1820	
			ESSENCE GINGER	1821	
76. P..............., S	1987	LONDON	MILK PUNCH	1823	
MAPLE SYRUP	1987		Silver gilt labels		
77. PEARSON, GEORGE	1817-1825	LONDON	84. RAWLINGS, CHARLES and	1829-1897	LONDON
BRANDY	c1817		SUMMERS, WILLIAM		
			ANISEED	1829	
78. PETERSEN, B & Co	1900	CHRISTCHURCH	LEMON	1829	
ANGELICA	c1900		COGNAC	c1830	
C. PORT	c1900		EAU DE ROSE	1830*	
			ESS. GINGER	c1830	
79. PHIPPS, THOMAS and	1783-1811	LONDON	BRANDY	1831	
ROBINSON, EDWARD			EAU DE COLOGNE	1831	
ELDER	c1783		EAU DE PORTUGAL	1831	
CLARY	1791		LAVENDER WATER	1831	
LEMON	c1795		ESCHALOTTE	1832	
RASPBERRY	1797		ROSE WATER	1836	
RATIFIA	1800		LAVENDER WATER	1836	
Milk of Roses	1808		DISTILLED WATER	1838	
			GRAPE	1839	
80. PHIPPS, ROBINSON & PHIPPS	1811-1816	LONDON	B (for BLACK INK)	1840	
BRISTOL WATER	1811		ARISETTE	1844	
LAVENDER	1812		WATER	c1844	
ROSE	1812		ANISETTE	c1845	
			NOYEAU	1856	

145

NAME	DATES	LOCATION	NAME	DATES	LOCATION
BRANDY	1860		RAISIN	c1800	
Silver gilt labels					
*Attributed		90. ROBB & WHITTET	1855-1880	EDINBURGH
			SWEET	c1860	
85. REILY, JOHN	1801-1826	LONDON			
LIME JUICE	c1801	91. ROBERTSON, WILLIAM	1795-1805	EDINBURGH
CURRANT	1802		SOURING	c1800	
MILK PUNCH	1805				
FRONTIGNIA	1813	92. ROBINS, JOHN	1774-1801	LONDON
COGNAC	c1815		NOYEAU	1796	
EAU DE COLOGNE	c1815				
ESSENCE OF ROSE	c1815	93. ROSSO, PAULO	1857-1870	MALTA
LAVENDER	c1815		BITTERS	c1860	
JESSAMINE	c1820				
PORTUGAL	c1820	94. SMITH, R. W.	1818-1850	DUBLIN
RAISIN	1823		FRONTIGNAC	c1818	
NOYEAU	1825				
LEMON	c1825	95. SNATT, JOSIAH	1797-1817	LONDON
Silver gilt labels			SAL VOLATILE	1801	
			MUSCATE	1804	
86. REILY, MARY and CHARLES	1826-1829	LONDON	P.VINEGAR	1804	
NOYEAU	1826				
RATAFIA	1827	96. STEWART, ALEXANDER	1820-1830	INVERNESS
VINEGAR	c1828		MILK PUNCH	c1820	
87. REILY, CHARLES and	1828-1855	LONDON			
STORER, GEORGE		97. STOKES and IRELAND	1878-1925	BIRMINGHAM
CAMPH DROPS	c1830		FRENCH	1924	
CAMPH.TD SPIRIT	c1830				
EAU D'ALIBURGH	c1830	98. STORR, PAUL	1800-1838	LONDON
ESS GINGER	c1830		KIRSCHENWASSER	1835	
ESSENCE OF GINGER	c1830		NOYAU	1835	
KIRSCHWASSER	c1830				
NOYAU	c1830	99. STREETIN, THOMAS	1794-1830	LONDON
POISON LAUDANUM	c1830		EAU DE COL	1829	
SPIRIT LAVENDER	c1830		TOOTHMIX	1829	
TOOTH MIXTURE	c1830				
GRAPE	1839	100. SUMMERS, WILLIAM	1859-1887	LONDON
			CRÈME DE VANILLE	1859	
88. RICH, JOHN	1765-1810	LONDON	VIOLET	1859	
Benedict	1802		LA ROSE	1871	
			COGNAC	1877	
89. RICHARDSON, RICHARD IV	1779-1822	CHESTER	TOILET VINEGAR	1885	
FRONTINIAC	c1800		EAU DE COLOGNE	1887	
MADEIRA	c1800		LAVENDER WATER	1887	

146

SILVER LABELS

NAME	DATES	LOCATION	NAME	DATES	LOCATION
101. TAYLOR, JOSEPH	1780-1827	BIRMINGHAM	CURRANT	1811	
KIRSHEN WASSER	1824		H	1811	
			LW	c1811*	
102. THORNTON, HENRY	1885-1912	LONDON	EAU D'OR	1814	
GINGER BRANDY	1895		CURRANT	1816	
BENEDICTINE	1895		MILK PUNCH	1820	
CHERRY BRANDY	1895		CORDIAL GIN	1830	
MARASCHINO	1895		PUNCH	1830	
			HUNGARY WATER	1835	
103. TIFFANY, CHARLES	1891-1902	NEW YORK	SHERBET	1839	
GUM SYRUP	c1895				
			112. YAPP & WOODWARD	1844-1874	BIRMINGHAM
104. TOLEKEN, JOHN	1795-1836	CORK	OIL	1850	
COLARES	c1795				
PARFAITE AMOUR	c1795				

* Unmarked. Attribution based on similarity to H, marked for 1811.

NAME	DATES	LOCATION
105. TROBY, WILLIAM BAMFORTH	1804-1829	LONDON
CORDIAL	1828	
106. TWEMLOW, WILLIAM	1790-1805	CHESTER
FRUIT	1799	
GINGER	1799	
107. UNITE, GEORGE	1824-1928	BIRMINGHAM
NOYAU	1827	
GINGER	1853	
APRICOT	1910	
BLACKBERRY	1910	
BLACKCURRANT	1910	
GREENGAGE	1910	
RASPBERRY	1910	
108. UNKNOWN MAKER	1900	BIRMINGHAM
VIOLETTE DE PARME	1900	
WHITE ROSE	1900	
109. WAKELY and WHEELER	1930-1960	LONDON
Silver gilt labels		
110. WATSON, THOMAS	1801-1811	SHEFFIELD
RASBERRY	1806	
111. WILLMORE, JOSEPH	1806-1845	BIRMINGHAM
VINEGAR	1808	

Fig 11.1

CHAPTER 11

MATERIALS

The attractiveness of boudoir labels is brought out in displays. Silver-gilt, porcelain and especially painted enamels lend themselves to beautiful presentations. The use of Sheffield Plate led to cheaper production. The portrait opposite (fig.11.1) is of Thomas Boulsover who discovered the methodology of producing it.

11.1 MARKED SILVER-GILT

Marked silver-gilt labels are few in number. Small size and lightweight articles may not have required to be hallmarked.

The following titles are included, not surprisingly they mainly relate to perfumes:

BERGAMOT	FLEUR D'ORANGE	LAVENDER WATER (2)
CRAB APPLE	FONTINIAC	MEAD
EAU DE COLOGNE (2)	FRANGIPANI	NOYAU
EAU DE ROSE	HUILE	NOYEAU
ELDERBERRY	HUNGARY WATER	PARFAIT AMOUR
ELDR. FLOR. WATER	JESSAMINE	PARSELY
ESPRIT DE ROSE	KIRCHEN WASSER	PORTUGAL
ESSENCE OF ROSE	LAVENDER	VINAIGRE

11.2 MARKED SILVER

An alphabetical list of some of the titles is set out below for the purposes of easy reference:

ACID	CAMPH DROPS	EAU DE COLOGNE (5)
ALBA FLORA	CAMPHORATED SPIRITS OF	EAU D'OR
ALBA-FLORA	WINE	EAU DE PORTUGAL(2)
ANGELICA	CAMPH.TD SPIRIT	EAU DE ROSE
ANISEED	CHARTREUSE	EAU DE TOILETTE VERVEINE
ANISETTE	CHERRY (3)	ELDER (2)
ANNISCETTE	CHILLI VINEGAR	ELDER FLOWER
APPLE	CLARY	ESCHALOTTE
APRICOT	COGNAC (3)	ESPRIT DE ROSE
ARBAFLOR	COLARES (2)	ESS GINGER
ARISETTE	CORDIAL	ESSENCE (2)
AROMATIC VINEGAR	CORDIAL GIN	ESSENCE GINGER
ARQUEBUSADE	COWSLIP	ESSENCE OF GINGER
ARQUEBUZADE	Crème de Noyeau	Eyes
ASTRINGENT	Crème de Thé	FONTANIAC
AURQUEBUSADE	CRÈME DE VANILLE	FRENCH (2)
B	CURACAU	FRENCH VINEGAR
BALM	CURRANT (8)	FRONTIGNAC
Benedict	CURRANTS	FRONTIGNAN
BITTERS	DAMSON	FRONTIGNIA
BLACKBERRY	DANDELION	FRONTINIAC
BLACKCURRANT	DENTIFRICE WATER	FRUIT
BRANDY (8)	DISTILLED WATER	GINGER (7)
BRISTOL WATER	EAU D'ALIBURGH	GINGER ESSENCE
C. PORT	EAU DE COL	GOOSEBERRY

MATERIALS

GRAPE (3)
GRAPEFRUIT
GREENGAGE
GUJAVA
GUM SYRUP
H
HAIR WATER
Hartshorn
HUILE
HUILLE DE ROYAUX
JESSAMINE
KIRCHENWASSER
KIRSCHENWASSER
KIRSCHWASSER
KIRSHWASSER
LA ROSE
LAVENDER
LAVENDER WATER
LEMON (4)
LIME JUICE
MADEIRA
MAPLE SYRUP
METHYLATED (2)
METHYLATED SPIRIT

MILK PUNCH (3)
MILLE FLUERS
Milk of Roses
MUSCATE
NECTARINE
NOYAU (3)
NOYEAU (7)
ODC
ODV
OIL (3)
ORANGE (3)
ORANGE BITTERS
ORANGE CURACAO
PARFAITE AMOUR
PARSNIP
PEPPERMINT
POISON LAUDANUM
PORTUGAL
P.VINEGAR
RAISIN (6)
RASBERRY
RASPBERRY
R.CONSTANTIA
RATAFIA

RATIFIA
RATITIA
RATTAFIER
ROSE
ROSEWATER
S. WATER (2)
SAL VOLATILE
SHERBET
SOURING
SPIRIT LAVENDER
SWEET
TISSANNE
TOILET VINEGAR
TOOTH MIXTURE
TOOTHMIX
VERBENA
VINAIGRE
VINEGAR (3)
VIOLET
VIOLETTE
VIOLETTE DE PARME
WATER
W.CONSTANTIA
WHITE ROSE

11.3 UNMARKED SILVER AND SILVER-GILT

Silver-gilt was used for small perfume labels adorning scent bottles. Silver was also used by some hundred or so makers of boudoir labels. Many items of silver or looking and feeling like silver are unmarked. There are a number of American sterling silver labels (standard .925) such as those recorded for AMONIA, COLORLESS HAIR TONIC,NEW MOWN HAY and TOILET WATER. The titles CAMPHOR JULEP, EAU DE COLOGNE and ROSE WATER have also been noted, all in silver-gilt, and all unmarked. Some of these unmarked labels are recorded in the list below alphabetically for purposes of easy reference:

ACQUA D'ORO
ATTAR OF ROSES
AMMONIA
BERGAMOT
BLACK
CALAMITY WATER
CAMPHOR JULEP
CINNAMON
CITRON
CITRONELLE
CLOVES
COPYING
CRÈME de CAFÉ
CRÈME de NOYEAU ROUGE
Delicieux
EAU DE COLOGNE
EAU DE COLODON
ELIXIR
ELIXIR DE SPA

ESS.BOUQUET
ESS BOUQUET
F. CREAM
FOUR BORO'
GRANDE MAISON
HUILE DE VANILE
HUILE DE VENUS
HUILE DE VINUS
JASMIN
JONQUIL
LAUDANUM
LAVENDAR
LAVENDER
LAVENDER
LAVENDER WATER
LAVENDER WATER
L.W.
MALVERN WATER
MINT

NOYAU
OIL
OPPOPONAX
POISON
RED
RED VINEGAR
RONDELETIA
ROSE
Rose
ROSEMARY
ROSE WATER
ROSE.WATER
ROSEWATER
SELTERS
SELTZER WATER
SEVE
Star Mise
VINEGAR
VIOLET
WOODBINE

11.4 SILVER PLATE

There are a number of Old Sheffield Plate labels such as those recorded for ALBA FLOR, CURRANT, GINGER, GOOSEBERRIES, NOYEO, RASPBERRY and SULTANA and of copperback escutcheons such as that recorded for RAISON, and of Continental Plate such as that recorded for Fr. d'ORANGE. A set of plated labels for RAISIN, GOOSEBERRY and ORANGE is illustrated in fig. 11.2 and for CHERRY, ELDER and CURRANT in fig. 11.3. Some of the surviving electroplated labels and labels of various other similar forms of silver plated base metals are noted alphabetically below for the purposes of easy reference:

B, CURRANT	GINGER CORDIAL	MALVERN WATER
BITTERS	GINGER WINE	MELOMEL
BORACIC WATER	GLYCO-THYMOLINE	MILK PUNCH
C. De ROSE	HUILE DE VANILLE	ORANGE
CHERRY	LACHRYMA	PONDS EXTRACT
CITRATE MAGNESIA	LAVENDER WATER	R. VINEGAR
CURRANT	LIMOSIN	RAISIN
Drinking Water	LISTERINE	RATTFIA
EAU DE NUIT	LOVAGE	SPRUCE
FLORIDA WATER	LYSTERINE	TANGERINE

Fig 11.2

Fig 11.3

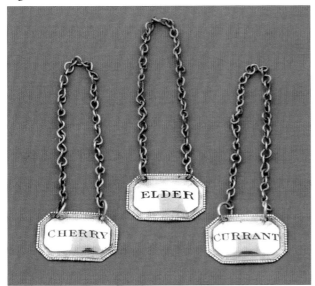

Fig 1.5

11.5 BONE INCLUDING IVORY

There is an interesting VINEGAR label made out of bone which looks inappropriate for use on the oil and vinegar frame or soy frame. VINEGAR was used for a variety of purposes. One use which concerned the boudoir was the use of AROMATIC VINEGAR to overcome a fainting fit. However the ivory VINEGAR could have formed part of the sauce label group of other ivory rounded rectangulars inscribed for CAYENNE, HARVEY, PRATTS, QUORNDON and WORCESTER. Ivory was used for NOYAU and VANILLE, MOSELLE & HOCK (see fig. 11.8). There is also a small label in ivory for PORT with a long chain, perhaps for a boudoir decanter. Small ivory collars have been noted for GINGER (see fig. 9.126b). Tiger claws speak for themselves (see fig. 1.5)

Fig 9.126b

11.6 MOTHER-OF-PEARL

Two crescent shaped labels have been noted. One is for ACID which has a magnificent border of twelve six-pointed stars. The other is for LOCH LOMOND,

Fig 9.1

an interesting perfume. A small mother-of-pearl label has been noted for GRAPE. A medium size for GINGER has been illustrated in the Journal (11.1). Another for EAU DE NOYAUX has been mentioned in the Journal (11.2). Another for NOIEAU was for use in a travelling decanter service (11.3). KIROCHWASSER is a title appearing on a mother-of-pearl label and see fig. 11.9.

11.7 PEWTER

A set of four slot boudoir labels has been noted, for Carnation (fig. 9.50), Sweet Pea, Lily of the Valley and White Lilac, made in London around

Fig 9.50

1900. The slot labels were retailed by J. Floris and Co. They match similar slot labels made in silver in 1905, with ivory labels inserted, stamped FLORIS. Pewter labels are rare. An oval shaped RUM has been described in "Wine Labels". A WHISKY has been noted stamped "TINN" and also stamped with a capital letter H in a shield. A set of six subdued narrowish boudoir tipple escutcheons in pewter with engraved scrolling designs has been noted (11.4) for BOURBON, BRANDY, GIN, SCOTCH, SHERRY and VODKA. Pewter was not regarded as elegant enough for the dining room.

11.8 ENAMEL

Fashionable London in the 1740s encouraged English enamellers to include labels in their productions. They had already had great success with enamel dials for watches and clocks since the early 1700s. By 1740 these had become quite elaborate and enamelling was being widely practiced. Production of boudoir labels was given a boost by the success of enamelled necessaires, enamelled scent bottles and the work of French artists. Painting with enamels was leading to greater artistic freedom. A boost to the production of English enamels was given by the new manufacture at Battersea. Copper plates were used as noted by Horace Walpole in 1755. Copper based enamels were affordable because the development of transfer printing by John Brooks in Birmingham from 1751 aided mass production. Simon Ravenet's connection with the firm of Janssen, Delamain and Brooks of York House Battersea began in 1753. His work was in monochrome, in delicate shades of red, puce, brown or mauve, or in black. According to the sale advertisement in 1756 placed in the Daily Advertiser he made "Bottle Tickets with Chains for all sorts of Liquor". Examples of his work include HUILE DE VENUS, MUSCAT, PUNCH, RAISIN, ALE, CURRANT and QUETSCHE, all of which might have been used in the boudoir.

Enamel labels were very popular in the boudoir. As a very general rule of thumb wine labels showed grapes and boudoir labels showed flowers (figs. 11.5 and 11.6). Samson enamels were made by the firm of Samson, Edme et Cie of Montreuil, Seine-Saint-Denis in France until 1959. The firm was set up in 1845 by Emile Samson's father in the Rue Vendome. It specialised in copying productions of other factories such as those established in Bilston, Birmingham, Dudley, Liverpool, London, Wednesbury and Wolverhampton. Two painted escutcheon enamel labels, formerly in the Weed Collection, for CRÈME DE MENTHE and RUM, belong to the English 1760s group. It is difficult to attribute the precise source of manufacture. John Sadler and Guy Green operated the Liverpool Printed Ware Manufactory in Harrington Street from about 1749, claiming indeed that they had invented

Fig 11.5

Fig 11.6

transfer printing. Between Birmingham and Bilston there were at least 40 manufactories. Yardley's of Church Hill Wednesbury were well known during the period 1776 to 1859. Dovey Hawkesford of Bilston were active from 1749. Benjamin and John Bickley were well known enamellers from Ettinshall Lodge.

Difficulties in attribution arise from the fact that Samson labels were such good reproductions! The crossed SS mark attributed to Samson was rarely applied. Only some five examples have been noted. The following indicative but not definitive tests for classifying an enamel label as a Samson label have been devised:-

1. Does the shape of the label match the shape of a similar but blank label in the custody of the Victoria and Albert Museum?
2. Is the type of attached chain similar to the chain on figures 1090 and 1091 illustrated in "Wine Labels"?
3. Does the style of decoration match that shown on figure 1089 illustrated in "Wine Labels"?
4. Is the capital letter "I" if any in the title surmounted by a dot?
5. Are serifs prominent and the letters thin stroke followed by thick stroke in the lettering of the title?
6. Does the label show fluorescence?
7. Is the painting raised and similar to known examples?
8. Are the flowers popular French varieties?
9. Is green copper oxidisation apparent?
10. Is the metallurgy test consistent?

Results from the application of these tests show that many labels in the past attributed to England should in fact have been attributed to Samson of France.

Enamel labels come in two basic sizes in width, 3.6cm to 3.9cm for small labels and 5.1 to 5.2cm for large labels. Six small labels are illustrated. BORACIC, ELDER FLOWER WATER and the two examples of METHYLATED SPIRIT are Samson, with Samson chains and a blue tinge to the edges of the enamel backing. Later Samson reproductions have a tendency for green copper oxidisation to seep into the edges of the white enamel backing. EAU DE TOILETTE has a different style chain. BORIC ACID is quite different. It has a stronger colouring and is marked on the reverse "FRANCE" in red. Four large labels are illustrated (fig. 11.5). MOUTH WASH, EAU DE COLOGNE and BATH POWDER are all probably Samson. ANISETTE, VERVEINE and JASMIN are French but not Samson. They measure 5.1cm in width and are marked "Dreyfous" on the reverse. So also KIRSCH and BRANDY are French but not Samson. They are marked "Delvaux" on the reverse. INDIAN is also 5.1cm in width, and may be by Dreyfous.

Wines and spirits were drunk in the boudoir. The enamel labels used in the boudoir were appropriately often decorated with roses instead of grapes and vine leaves. Exceptions include HUILE D'ANIS DES INDIES, HUILE D'ANIS ROUGE, and CRÈME DE PORTUGALE. The enamels were delightfully decorative and informal - see the Commentary. One can see illustrated in Dr. Wells' "Samson Enamel Wine Labels" at pages two to five CLARET matching EAU DE TOILETTE, PORT matching CHERRY-BRANDY, PORT matching EAU DE LUBIN, WHISKEY matching TONIC, CHARTREUSE matching

VIOLET and SLOE GIN matching WHITE ROSE in terms of design. PORT in fact was a tipple much favoured by the ladies. FINE NAPOLEON and VIELLE FINE were drunk to settle the stomach. Some of these labels are listed alphabetically below:

ALCOHOL	EAU DE NOYEAU	Limonade
ALKOHOL	EAU D'ORANGE	Limoux
AMMONIA	EAU DE ROSE	MARMALADE
ANISETTE	EAU DE TOILETTE	METHYLATED SPIRIT
BATH ESSENCE	EAU DE VIOLETTE	MON PARFUM
BATH POWDER	ELDER FLOWER	MOSS ROSE
BATH SALTS	ELDER FLOWER WATE	MOUTHWASH
BENZIN	ELDER FLOWER WATER	MUSCAT
BENZOIN LOTION	ELIXIR DENTIFRICE	ORANGE FLOWER
BENZOIN SALTS	EYE DROPS	PERFUME
BITTERS	FINE NAPOLEON	PEROXIDE
BORACIC	G. BEER	PLUM
BORIC ACID	GARGLE	Pt. George
BRANDY	GLICERINE	RAISIN
CASTOR OIL	HAMMAM BOUQUET	RASPAIL
CHERRY-BRANDY	HUILE D.	RASPBERRY
CHIPRE	HUILE D'ANIS DES INDES	Ratafiat de fleur d'orange
Cogniac	HUILE D'ANIS ROUGE	SAVRON
CORDIAL	Huile de Rose	SCUBAC
CRÈME DE PORTUGALE	HUILE DE VANILLE	STRAWBERRY
CRÈME DE VANILLE	HUILE DE VENUS	TOILET POWDER
DACTILLUS	HUILLE D'ORANGE	TONIC
EAU BORIGUEE	JASMIN	VERVEINE
EAU BORIQUEE	JASMINE	VIELLE FINE
EAU DE CELADON	KIRSCH	VIOLET
EAU DE COLOGNE	LAVANDE	WHITE LILAC
EAU DE LUBIN	LILAC	WHITE ROSE

The French SAVRON and BITTERS and the Battersea RAISIN were exhibited at the Circle's Spring Meeting in 2008. The following wine and sauce enamel labels were also exhibited at that time:

ALIUM TICO	LA COTE
ANDAYE	LE MUSIGNY
BOURGOGNE	MADEIRA
BURGUNDY	RHUM
CALABRE	ROTA
CANON	SEROS PIGALLE
CHAMPIGNON	Sherry
COTILLION	ST. PEREST
CURACAO	ST. PERET
DU RHIN	TOKIE
GRANDE CANAS	

11.9 PORCELAIN

Labels made of porcelain are in the main Crown Staffordshire productions (CS) following pattern number A6297, often retailed by T. Goode & Co. They are marked with a CROWN, a crown monogram, STAFFORDSHIRE and ENGLAND. The earliest example for "FRENCH" also includes "ESTB. 1801" either side of the crown and "MADE IN" in front of ENGLAND (fig. 11.7). The pair BORACIC LOTION and GLYCERINE also include "A.D. 1801" but this is positioned below ENGLAND. The titles of EAU DE COLOGNE and FRENCH are in gold lettering. They were made between 1892 and 1923. Other porcelains are French porcelains (FP) or products of the Samson factory (SF). One is a Coalport production of around 1932 in date (CP). IT IS MARKED "BONE CROWN R CHINA COALPORT MADE IN ENGLAND EST.1750" in blue and "24" in green. It has its original presentation box which is illustrated in fig. 9.91b. Bin labels are not included in the following list of some porcelain labels as they do not fall into the boudoir category.

Fig 11.7
Porcelain Labels

AMMONIA (CS) (1) (2) (3)
BATH POWDER (CS Group C) (2)
BENZOIN
BORACIC (CS Group A)
BORACIC ACID (CS)
BORACIC LOTION (CS) (1) (2) (3)
CACAO (FP)
Chloroform
CHYPRE (FP)
EAU DE ROSE   (CS Group B) ((3)
EAU DE COLOGNE (CS Group B) (1) (3)
EAU DE COLOGNE (FP)
Eau de Nuit (CP)
EAU FROIDE (FP)
EYE DROPS (CS)
EYE WASH (CS)
FRENCH (CS)
French (FP)
GIGGLEWATER
GLYCERINE (CS Group A) (1) (2) (3)

GOUTTES de MALTHE (FP)
Graine de lin   (FP)
INDIAN TEA (CS)
LAVENDER WATER (CS) (1) (3)
Limonade
LIQUID POWDER (CS) (1) (2) (3)
METHYLATED SPIRIT (CS) (1) (2) (3)
ORANGEADE
RATAFIA
Ratafiat De Fleur d'Orange (FP)
TOILET LOTION
Vinaigre Ordinaire (FP)
WHITE ROSE

Key to List of Porcelain labels set out above:
(1) Marked with the painter's initials "IL" in red.
(2) Overmarked "T. GOODE & Co. SOUTH AUDLEY STREET LONDON W".
(3) Marked with pattern number A6297 in red.

11.10   TORTOISESHELL

Labels made from the shell of the sea turtle, like tea caddies made of this material, are rare. They tend to be fitted with a chain made of the same material. Environmental restrictions make it difficult to repair damaged labels or chains. Only one boudoir label has been noted, having a chain of the same material and cut from one shell with its title painted in gold, and this was for:

CHERRY.

11.11   GLASS

Labels made from glass or a glass-like substance are rare. No boudoir labels have been noted. However of general interest is the recent excavation of a Sixteenth Century Inn at Holborn Viaduct containing a number of wine bottles with labels containing owner's initials as part of the bottle and an on-site micro-brewery. According to the report (11.6) "The most interesting discovery is a glass medallion that would have hung round the neck of an onion-shaped bottle of wine. It features the pub's logo, a picture of three barrels, and is surrounded by the text "at the 3 Tuns at Holborne Bridge". It also featured letters thought to be the initials of the landlord and the landlady. Initialled wine bottles were taken to the local wine merchant for re-filling. Some silver labels bear owner's initials or armorials.

11.12   PLASTIC

A. H. Woodfull was best known for designing stylish plastic tableware in the 1950s. He designed many products for British Industrial plastics between 1934 and 1975 (11.7). His three piece beetleware cruet set in urea formaldehyde produced in 1946 did not involve the use of plastic sauce labels. No suspended plastic labels have been recorded. The brown and cream bakelite produced by toymakers Cowan De Groot & Co (CODEG) could have been a good substitute for tortoiseshell.

11.13   LEATHER

A gilt metal boudoir tipple for SHERRY with a border of oak tree leaves and acorns is set on tooled red leather with a double-reeded gold border; and see Cognac, fig. 11.8.

11.14   CARDBOARD

A set of three demountable cardboard labels on boxed decanters has been noted for Cooking Brandy, Damson Cordial and Ginger Wine, all having medicinal benefits. Cardboard or stiff paper titles were also made for silver slot labels by Britton, Gould & Co. in 1905 for Messrs . J. Floris.

Fig 11.8 Ivory Boudoir Labels and a large leather-backed Cognac

NOTES

(11.1)   See 12 WLCJ 1, p.18
(11.2)   3 WLCJ 6, p.133a
(11.3)   5 WLCJ 4, p.94
(11.4)   Country Life, 9.11 1978, page 1490
(11.5)   Featured on Antiques Road Trip shown on
20.10.2011; sold at auction in Shrewsbury for £30 hammer price.
(11.6)   Jack Malvern, The Times, 5.8.2011, p.9
(11.7)   The Times obituary, 28.6.2011, p.48

Fig. 11.9 Mother-of-Pearl Boudoir Labels

Fig 12.1

CHAPTER 12

BOUDOIR TIPPLES

12.1 These relate to names of alcoholic drinks, all of which are included in the Alphabetical Master List Of Names Of Wines, Spirits, Liqueurs and Alcoholic Cordials on Bottle Tickets and Bin Labels as at 31st January 2010, which were often displayed upon smaller sized labels than the average wine label, and could have appeared in the boudoir, perhaps being of feminine appeal in appearance, and being displayed on tall, slender, elegant decanters containing tipples Eau de Vie, Vin and Chartreuse (fig. 12.1) or on a novelty decanter (fig.12.2) or on tall and elegant narrow spirit decanters (fig.12.3) not suitable for the dining room. STOUT, for example, a refreshing boudoir tipple, is the title of a tiny double reeded octagonal label by Rawlings and Summers, made in 1862 (fig. 12.4). On the other hand ALE, a copper backed plated label is much larger and clearly not intended for the boudoir, although Ravenet's ALE might have been. The plated label might well have been used in the kitchen, cellar or storeroom. Boudoir tipples would be dispensed from decanters suitable for use in the boudoir. One such is illustrated (fig. 12.2), being a Victorian novelty silver mounted glass snake liqueur decanter made by Saunders and Shepherd right at the beginning in London of the art nouveau period in 1895. The design, which involved gilded highlights being applied to the snake's green glass body, was registered under number 260386. The novelty decanter, it seems, was introduced by Alexander Crichton in 1881 using animal subjects such as a cockatoo, a duck, a drake, an otter and a penguin. The glass was made in Stourbridge either by John Northwood or by Thomas Webb.

12.2 Wine making was a popular activity. Part of a typical hand-written household recipe book for household wines was compiled between

Fig 12.2

Fig 12.3

Fig 12.4

Fig 12.6

Fig 12.5

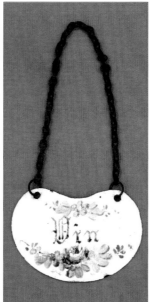

Fig 12.7

Fig 12.8

1740 and 1770 and has been reproduced in the Journal (12.1). It gives recipes for making APRICOT, BARLEY, BAUM, BIRCH and SYCAMORE, for example (part illustrated in fig. 6.6). The wine-making section lists some 156 recipes. There are also 16 recipes for RATAFIA, 82 for drams of various sorts, 24 for "dyet drinks", 16 for PUNCHES and 18 for VINEGARS. CANA, a red wine from Galilee, has been produced by Cremisan Cellars since 1885. The monastery of Cremisan is sandwiched to-day between some Israeli Apartment Blocks and some Palestinian luxury villas and is on the border between the West Bank and Jerusalem. The monks belong to the Silesian Order. The name CANA appears on a plated octagonal plain curved label with a heavy chain (fig. 12.5). A typical French tipple is illustrated being an enamel label for VIN-DE-CÉRISE or CHERRY WINE (fig.12.6). Home wine production is perhaps best marked by a simple boudoir label inscribed for Vin (fig. 12.7). A list of labels that might have been used in connection with home made wines is appended below.

12.3　With regard to liqueurs, these could have been dispensed as a "night cap" from toddies using a simple unmarked silver label reading "Liqueur" in lower case or script lettering (fig.12.8). Flavoured brandies such as CHERRY BRANDY and LIQUEUR BRANDY

(fig. 12.8.a, enamel, made in France it seems from the crossed device mark shown in fig. 12.8b) were much in vogue, especially as they had medicinal properties. CURACAO was very popular as evidenced by a variety of labels including a plain unmarked silver octagonal (fig. 12.9) and various enamels (figs. 12.10, 12.11 and 12.12) as was BENEDICTINE with its monastic connections (figs 12.13 and 12.14). The search for the ELIXIR of long life was begun by monks in the Middle Ages and continued in wealthy establishments. The person in charge of the still room had to have acquired a knowledge of essences, spirits, alcohols and distillations. Originally colour was an easy way to distinguish liqueurs from eaux and crèmes which were white, elixirs which were yellow and ratafias which, being fruit infusions, took on the colour of the fruit in question. The herb garden produce came in handy for flavouring, using such plants as ROSEMARY, ELDER, ROSE and BALM MINT. Some experimental liqueurs were made for home consumption, such as an example using the flowers of citron, and for the assistance of the sick. Some liqueurs were used in food preparation as a flavouring. Maraschino was used in the cooking of a lunch at Waddesdon for Queen Victoria when she visited there in 1890, according to an extant menu. A list of labels that might have been used in connection with liqueurs, many of which were originally medicinal productions, is appended below.

Fig 12.8a

Fig 12.8b

Fig 12.9

12.4 Cordials made one feel better. They were restoratives which aided relaxation. There were plenty of recipes. They were very much to hand in the boudoir. Cordials used for medicinal purposes were also made, such as GOLD WATER or GULDEWATER following, for example, the recipe of Arnold de Villeneuve which he says "is very different from wine both in colour and in substance, in effect and in operation". One difference was certainly made by the addition of macerated herbs, aromatic spices and even minute quantities of metallic gold. A list of labels that might have been used in connection with alcoholic cordials is appended below.

Fig 12.10

Fig 12.11

Fig 12.12

Fig 12.13

Fig 12.14

Fig 12.15

Fig 12.16

Fig 12.17

12.5 Cocktails also made one feel better. BITTERS (fig. 9.32) were even licensed for import in bottles from the United Kingdom into the United States in time of prohibition (1922-1923). KHOOSH BITTERS according to the Chanticleer Society, a worldwide organisation of cocktail enthusiasts, is a recorded title. The Society has noted an opened bottle with this title and that there is an unopened version in the Museum of the American Cocktail (12.2). An export label for KHOOSH is stamped on the reverse "Made in England" (fig. 9.149a). The art deco period provided a wonderful background for increasing the popularity of cocktails. French labels for BITTER (figs. 9.31a and 9.31b), VERMOUTH DE TURIN (fig. 12.15), Vermouth Rouge (fig.12.16) (made by Barker Ellis Silver Co. in Birmingham in 1965) and CHERRY BROUDY (fig.12.17) are illustrated. VERMOUTH of Turin was invented circa 1786 by Antonio Benedetto Carpano who came from Turin (1764-1815). It was made from white wine which had been added to an infusion of some thirty different herbs and spices. It was then sweetened with a little spirit. It was regarded as a boudoir drink being held suitable for ladies! Then it became an ingredient for the making of some cocktails. The cocktail bar and the cocktail shaker provided the essential equipment and support for the cocktail. Shakers were sometimes made of chromium plate and bakelite plastic. To assist the barman recipes were incorporated such as in the English made "INCOLOR" produced for the American market (patent said to be pending) where rotating the top section revealed a recipe for "Bronx" being one-third dry GIN, one-third dry VERMOUTH and one-third orange juice all with ice and a recipe for "Clover Club" being two-thirds dry GIN, one egg white, a dash of GRENADINE and one-third LEMON juice all with ice. After mixing a strainer was incorporated to deal with objects like pips, mint and peel. Finally at the base was a measure in gills. A list of cocktail titles compiled by an American enthusiast is set out in the Cocktail Chronicles (12.4).

12.6 With regard to spirits, it is thought that IRISH WHISKEY was produced as long ago as the Twelfth Century by Irish monks. The famous Old Bushmills Single Malt Distilery in County Antrim was licensed in 1608 during the reign of King James I. Water from the River Bush was mixed with barley and fermented. It was then triple distilled unlike SCOTCH WHISKY which was double distilled. Then it was matured in re-used oak barrels for the

Fig 12.18

legal minimum of at least three years. The maturation period could, however, be as long as forty years. Some oak barrels came from Spain and Portugal where they had been used to store and give flavour to SHERRY and MADEIRA. Other oak barrels came from Kentucky in the USA where they had been used to store BOURBON. After the maturation process came the blending process. The WHISKEY was drained from the barrels, mixed up and then stored in large vats prior to bottling. Domestic distilleries flourished in the UK from the time of King William III of Orange, especially as regards the making of GIN so much liked by the ladies (fig. 12.18). Illustrated is a splendid art deco enamelled boudoir GIN set in a silver mounting by Turner and Simpson in 1934 (fig. 12.19). Also illustrated is a porcelain art deco boudoir RYE (fig. 12.20) and an enamel art deco boudoir KIRSCH (fig. 12.21) with so-called stove enamel backing marked by Delvaux of 18 Rue Royale in Paris (fig. 12.22). PLYMOUTH GIN as made in the Blackfriars' Distillery from 1793 gained its particular flavour from soft Dartmoor water to which was added juniper, lemon, orange, angelica and orris amongst other flavourings. The Brewers' Livery Company in the City of London is fortunate to have received a recent Master's Gift of a silver-plated trolley in the form of a four-wheeled brewers' dray (similar to those in daily use by Wadworth's Brewery in Devizes in Wiltshire) carrying three identical banded barrels each with a spigot and bucket for the boudoir tipples being tots, noggins, shots, snifters, toddies, totties or the like. The trolley bears a design registration mark for 18th April 1863. Each identical barrel would have had its contents for the time being identified by a label suspended from the plug, but unfortuantely the labels are missing. Tots of the spirits WHISKEY (fig.12.23 containing examples by Heath and Middleton in 1904, Samuel Jacob & Co. in 1911 and Hukin and Heath in 1929) and BRANDY and of the liqueurs CHERRY BRANDY, CRÈME DE MENTHE (fig.12.24 Hukin and Heath 1924)and even KUMMEL (fig.12.23 Hukin and Heath 1924) were popular imbitions from mid-Victorian times. They were thought to be a good night cap. These were dispersed from small silver mounted glass jugs or toddies (see Chapter Four above). They were adorned with small kidney shaped labels or sometimes a small narrow rectangular label with rounded ends (see fig.12.8). The last letter of the scripted title was usually given a tail in

Fig 12.19

Fig 12.20

Fig 12.21

Fig 12.22

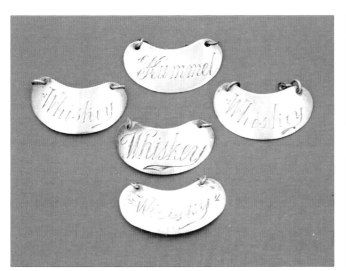

Fig 12.23

Fig 12.24

Fig 12.25

Fig 12.26

Fig 12.27

Fig 12.26a

the form of a scripted flourish. The toddies were refilled from a large glass storage jar. John Bird and Sons, glass makers of Glasgow, made a storage jar, inscribed "Fine Old Scotch", circa 1880, fitted with a spigot. A rather large green bottle containing 280cc, not incidentally designed for hanging on the wall, has been noted (12.3), said to be for WHISKEY tots or wee drams distributed under the direction of the local laird from the comfort of the Bar of the Fortingall Hotel located in the Arts and Crafts village of that name near Aberfeldy. The large bottle itself was adorned with a curved long narrow rectangular silver label, decorated with a beaded edge and suspended by a silver chain, made by Hubert Thornhill of Bond Street London and fully marked for 1888. It bears the mark he entered as a small worker on 14th July 1887. He was a member of the firm of Walter Thornhill & Co. The bottle bears a silver mount with a lockable cap, the laird's initials and the date of May 1888. It functioned therefore as a kind of tantalus. Even more unusual is the title engraved on the silver label which appears to be something like "SKARRAG AN T-SEAMH UASHITARAIN". This surely must be the longest recorded title! RUM (figs.12.25 made in 1802 by Phipps and Robinson and 12.26 bearing a maker's mark only for either PT or FT) and in Holland RUMM (fig. 12.26a, unmarked) and in Scotland SHRUB (figs. 12.27 marked with maker's mark JP only and 12.28 by Robert Gray and Son of Glasgow marked (see fig. 12.29) in 1815 in Edinburgh were also favourite tipples.

12.7   The decanter travelling box for use in various locations generally housed four decanters and eight drinking glasses, of a size suitable for cordials, liqueurs or spirits. A French boxed four bottle decanter frame shown on the Television programme "Flog it!" had a central ring handle for lifting it out of its box and was accompanied by four liqueur glasses matching the decanters. It was very

Fig 12.28

Fig 12.29

Fig 12.30

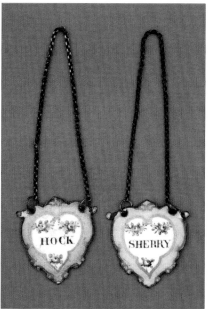

Fig 12.32

BRANDY (unmarked shield shape, fig. 12.35). For the Gentleman's retiring room one might expect to receive MADEIRA, SHERRY or CLARET (fig. 1.5). Spirits with a very high alcoholic content are represented by LIKOR and SNAPS from Denmark or Sweden in the art deco period with enamel filled titles (fig. 12.36) and by the quartet of Saides, Renadt, Gouffin and Danski (fig.12.37). Enamels were often used for spirits, such as in the case of FRAISETTE (fig.12.38) and FLORENCE M (fig. 12.39). A list of labels that might have been used in relation to the drinking of spirits and fortified drinks is appended below.

Fig 12.33

discreet and, for example, could easily be fitted into the Green Boudoir at Waddesdon amongst the French furniture. Typical boudoir decanter sets are represented by PORT, GIN and WHISKY

Fig 12.31   (made of porcelain, fig. 12.30), PORT, SHERRY and BRANDY (made of enamel, fig.12.31), HOCK and SHERRY (made of enamel, fig.12.32 and marked with a crossed device, fig.12.33), SHERRY and BRANDY (fig. 12.34) made of Crown Staffordshire porcelain marked England Pattern no. 6297 (BRANDY) and Made in England Pattern no. A6297 (SHERRY) and SHERRY, HOLLANDS and

12.8   Other alcoholic drinks possibly made at home are listed. The home cider press was not uncommon in country homes and farmhouses in the West Country. A French enamel for CIDER is illustrated in fig. 12.40. The micro brewery may well have been installed in farms. One has been discovered by archaeologists in a London pub, The Three Tuns at Holborn Bridge. Dr. Thompson, Physician to Frederick Prince of Wales, was advised by his friend Mr. Whitehead to "forbear to haunt cook-shops, hedge-alehouses, cider-cellars, &c"(12.5) which admonition paints its own picture. George Washington in 1757 wrote out by hand a recipe for Small Beer: "Take a large Sifter full of Bran Hops to your Taste. Boil these 3 hours then strain out 30 Gall into a cooler (and) put in 3 Gall Molasses while the Beer is Scalding hot. Let this stand till it is little more than Blood warm then put in a quart of Yeast if the Weather is very Cold, (then) cover it over with a Blanket & let it Work in the Cooler 24 hours then put it into the Cask"(12.6).

12.9   Home made wines perhaps tasted better having been decanted. Boudoir decanters were elegant, being tall and slender bearing small enamel labels such as GREEN CHARTREUSE, KUMMEL and FRENCH VERMOUTH or, as shown in fig.12.1, Chartreuse, Vin and Eau de vie, in typical boudoir style. Liqueur decanter frames on the whole tended to be slender or narrower and more graceful than for

Fig 12.34

Fig 12.35

Fig 12.38

Fig 12.39

Fig 12.36

Fig 12.37

Fig 12.40

wines. Spirits were dispensed not only from individual toddies but also from artistically decorated glass decanters housed in silver mounts (fig.12.3). Cocktails were dispensed from a cocktail bar and are a specialised subject (12.7). BITTERS has been noted housed in an attractive decorated glass bottle (fig. 9.178). Cordials were often poured from a labelled jug or decanter (see fig.12.41 for CORDIAL, an example by Robert Gray (fig.12.42) of Glasgow).

12.10 With regard to other alcoholic drinks not made at home, the 6th Duke of Portland had an elegant Beer Trolley which used to be trundled round Welbeck Abbey to provide refreshment where needed (12.8). Arthur Guinness so believed in his STOUT (fig. 12.4) that he signed in 1759 in Dublin a counterpart lease for 900 years. Guinness was recommended by some General Practitioners as a health giving tonic and even the White Rabbit has been said to have looked at his large watch and exclaimed "Oh, my ears and whiskers - it's Guinness Time!"

12.11 The labels included in the lists of Boudoir Tipples set out below are generally smaller than normal wine label size and are more informal, thus suggesting non-dining room use.

12.12 It may be appropriate to conclude this section on Boudoir Tipples with a tradtional toast:
    "There are tall ships,
    And there are long ships,
    There are ships that sail the seas,
    But the best ships
    Are friendships,
    And may they always be."

Fig 12.41

Fig 12.42

BOUDOIR TIPPLES

A. LIST OF SOME MAINLY HOME MADE WINES

APRICOT
BUDOCK
CALCAVELLA
CANA
CAPE
Champagne
CLARET
CONSTANIA
CONSTANTIA
COWSLIP
COWSLIP WINE
CURRANT
CURRANT WINE
CURRANT-WINE
DAMSON
DANDELION
DANDYLION
ELDER
ELDER FLOWER WINE
ELDER.FLOWER.WINE
ELDER WINE
ELDERBERRY WINE
GINGER WINE
GOOSBERRY
GOOSBERY
GOOSEBERRY WINE
GOOSEBERRIE
GOOSEBERRIES
GOOSEBERRY
GOSEBERY
GOUSEBERY
GRAPE
HOCK
M. WINE
MADEHERE
MADEIRA
MADE WINE
MOSELLE
MUSCAT
PARSNIP
PASS THE BOTTLE
PAXARETTE
PEPPER WINE
PEPPERMINT
P.MINT
RAISEN
RAISIN
RAISIN W.
RAISIN WINE
R.CONSTANTIA
RED CONSTANTIA
RED CURRANT
RED HOCK
STISTED
SWEET
Vin
VIN-DE-CERISE

W. CONSTANTIA
W. CURRANT
W.WINE
WHITE CURRANT

B. LIST OF SOME ALCOHOLIC CORDIALS

AMEIXORIA
ANIS
ANISEED
ANISETTE
ANNESEED
ANNISCETTE
ANNISEED
AQUA MIRABEL
BLACKBERRY
BLACKCURRANT GIN
BOUNCE
CHERRY
CINNAMON
CITROEN
CITRON
CORDIAL
CORDIAL GIN
CORDIALS
DAMSON CORDIAL
EAU D'ANDAILLE
FINE CITRONELLE
GINGER
GINGER CORDIAL
KIRCHEN WASSER
KIRSCHENWASSER
KIRSHWASSER
LAVANDE
LEDSAM VALE
MELOMEL
NOYEO
ORANGE
PEACH
PICK-ME-UP
QUETSCHE
RAISIN
RASBERRY
RASPBERRY
RATAFIA
RATAFIAT
Ratafiat De Fleur d'Orange
RATIFIER
RATTAFIER
SHRUB
STRAWBERRY
TANGERINE

C. LIST OF SOME LIQUEURS

ABSINTHE
ALLASCH
ANGELICA
ANISETTE

APRICOT BRANDY
BENEDICTINE
BOUNCE
CARLOWITZ
CASSES
CASSIS
CASSIS BLANC
Chartreuse
CHARTREUSE
CHARTREUSE GREEN
CHARTREUSE YELLOW
CHERRY BRANDY
Cherry Brandy
CHERRY-BRANDY
CHERRY BROUDY
CHERRY BY
CHERRY RATAFIA
CRÈME D'ALLASCH
CRÈME DE MENTHE
COINTREAU
CURACAO
CURACAO D'HOLde
CURACOA
DAMSON GIN
EAU D'OR
FRENCH VERMOUTH
Gd Marnier
GINGER
GINGER BRANDY
GINGER BRNDY
GOLDEN TRASSER
GOLD WASSER
GREEN CHARTREUSE
GULDEWATER
IT VERMOUTH
KIRSCH
KIRSCHENWASSER
KUMEL
KUMMEL
LIKOR
Liqueur
LIQUEUR BRANDY
LIQUEUR DE BIRSE
LIMONCELLO
MARASc
MARASCHINO
MINTE
MIRABELLA
MIRABELLE
MUM
NOYEAU
ORANGE BITTERS
ORANGE BRANDY
ORANGE CURACAO
ORANGE LIQUER
P. BRANDY
PARFAIT D'AMOUR
PEPPERMINT
PLUM
PLUM BRANDY

BOUDOIR TIPPLES

PLUM RUM
RASPBERRY B
RATAFIA
TRAPPISTINE
VERMOUTH
VERMOUTH DE TURIN
Vermouth Rouge

D. LIST OF SOME SPIRITS
AND FORTIFIED DRINKS

ABSINTH
ABSINTHE
AQUAVIT
AQUAVITAE
AQUAVITE
ARMAGNAC
BANYLS
BITTER
BITTERS
BOURBON
BRANDY
Brandy
CALABRE
CALVADOS
CANDIA W.
CANELLE
CANNELLA
Cherry Brandy
COGNAC
Cognac
Cogniac
Danski
DARK SHERRY
DESSERT SHERRY
DUNVILLE
EAU DE NOYAU
EAU DE NOYEAU
EAU DE VIE
Eau de vie
ESCALOTTE
FINE NAPOLEON

FLORe
FLORENCE M
FRAISE
FRAISETTE
GENEVA
GIGGLEWATER
GIN
GOMME
Gouffin
GRAPE
GRAPPA
GUJAVA
GUM
GUM SYRUP
H-GIN
HOLLANDS
IRISH B
KHOOSH
L'EAU CHAREC
LEMON BRANDY
LIGHT SHERRY
Lisbon
LOVAGE
MADEIRA
MALMSEY
MALMSEY DRY
MALMSEY RICH
MILK PUNCH
MOURACHE
NOYAU
NOYEAU
ORANGE BITTERS
ORANGE GIN
PEACH BRANDY
PERRY
PISCO
PORT
RED, PORT
Renadt
RHUM
RHUM FACON
RUM

RUMM
RYE
SACKE
Saides
SCOTCH
SHERRY
Sherry
SHRUB
SKARRAG AN T-SEAMH
UASHITARAIN
SLOE GIN
SNAPS
VIELLE FINE
VODKA
WHISKEY
Whiskey
WHISKY
Whisky
WHITE PORT
W PORT

E. LIST OF SOME OTHER
ALCOHOLIC DRINKS

ALE
BEER
CIDER
C.PORT
CYDER
MUM
STOUT

F. MIXED SET

CHARTREUSE
SCOTCH
CLARET
MARASCHINO
BRANDY

NOTES

(12.1)   See 8 WLCJ 1.
(12.2)   See Drinks of the World by James Mew.
(12.3)   Antiques Roadshow, Sunday 10th April 2011. Hot tots, toddies or totties tended to be served in silver vessels made by Mappin and Webb, Heath and Middleton and Goldsmiths and Silversmiths Company, for example, in the period 1900-1930. Some toddies were silver mounted conical glass containers with stoppers or lids.
(12.4)   See www.cocktailchronicles.com/drinks-index/ and marksexauer.wordpress.com; thecocktaillovers.com; cocktailians.com and adashofbitters.

com; all last visited on 5.8.2011.
(12.5)   See Dr. Scott's "Cottage Physician",1826.
(12.6)   Recipe from Washington's "Notebook as a Virginia Colonel" published in 2011 by New York Public Library.
(12.7)   See the following websites: adashofbitters.com; cocktailians.com; coctailchronicles.com; the cocktaillovers.com; and marksexauer.wordpress. com.
(12.8)   Welbeck Abbey is near Worksop in North Nottinghamshire. See www.hurleygallery.co.uk/ index.

Fig 13.1

CHAPTER 13

THE SILVERSMITHS

This chapter gives short details about some of the makers of silver boudoir labels including the illustrious Paul Storr (fig. 13.1). Some of the heading dates given after the makers name are tentative. They represent a makers approximate working dates. Many silver boudoir labels carry maker's mark only.

13.1   JOSEPH ANGELL 1811-1848

He was a working silversmith, the son of a weaver, situated at 55 Compton Street in Clerkenwell, until his retirement in 1848.

13.2   HENRY ARCHER 1910-1920

He was a Sheffield silversmith who made in 1916 an attractive ORANGE BITTERS label.

13.3   ARMY AND NAVY CO-OPERATIVE SOCIETY LTD 1903 – 1929

Henry Lawson, the Company Secretary, entered a mark in five sizes on 28th November 1903. Before then, from its foundation in 1871, the company had used the mark of F. B. McCrea, its manager. It had a large Silver Department. The range of its wares can be appreciated by examining its 1907 catalogue which was re-published by David and Charles of Newton Abbott in 1969.

13.4   JOSEPH ASH I 1799 – 1818

This competent smallworker occupied premises off Butchers' Hall Lane in the city of London, well positioned for catching passing trade to the Cloth Fair, St Barts and Smithfield from the London Wall area. His son was apprenticed to Joseph Biggs who in turn had been apprenticed to Robert Barker who was related to Susanna Barker (1778 – 1793), a prolific label maker and is noted for good quality boxes.

13.5   JAMES ATKINS 1792-1815

Before starting up on his own in 1792, James Atkins spent a year in partnership with John Essex. He specialised in making small silver items and buckles in particular, operating from 12 Well Street in Cripplegate. He made a number of sauce labels. By

his will dated 10th March 1814 he left his whole estate to his only daughter Theodoria Ann Atkins, also a buckle-maker. Probate was obtained in November 1815.

13.6   SUSANNA BARKER 1778-1793

Susanna Barker worked from 16 Gutter Lane near Goldsmiths' Hall. It seems likely that she took over Margaret Binley's business in 1778. She was a distinguished label maker and was followed by Robert Barker in 1793. Her GINGER ESSENCE label of 1791 was in the Greville Tait Collection and has been illustrated in the Journal.

13.7   HESTER BATEMAN 1761 – 1790

John and Hester Bateman, with their six children (John, Letticia, Ann, Peter, William and Jonathan). They moved in 1747 into 106-108 Bunhill Row, which linked via Blue Anchor Alley Old Street to Chiswell Street in the City of London and was well positioned on the boundary of the city with the adjoining borough. Letticia married Richard Clarke in 1754. Hester became a widow in 1760 and took over the business which she ran successfully with the help of family members until her retirement in 1790. She was a prolific label maker. Her products were of undoubted quality. Hence the APPLE label may have been made from a cutting up of another piece of re-used silver bearing Hester's mark.

13.8   PETER AND ANN BATEMAN 1791-1799

In 1798 Peter and Ann made a travelling box label for BRANDY which was of great assistance to weary travellers.

13.9   PETER, ANN AND WILLIAM BATEMAN 1800 – 1805

In 1791 Peter and Jonathan took over the business from their mother Hester. John died of cancer in 1791 leaving everything to his wife Ann, who joined up with Peter in running the business. Early in 1800 they were joined by Ann's second son William, Ann then retired in 1805. Their CHERRY cordial label is marked for the silver year 1799 to 1800.

13.10   PETER AND WILLIAM BATEMAN I 1805 – 1815

During this period before Peter's retirement in 1815, William gained considerable stature as a silversmith, getting ready to join the Livery of the Goldsmith's Company in 1816, going on to be Prime Warden in 1836.

13.11   WILLIAM BATEMAN I 1815 – 1827

William took on his son William II as his apprentice in 1815 for seven years. In 1822, the year in which he made BALM, William II joined his father and entered his own mark in 1827 upon taking over the business from his father. At this time William I was about to join the Court of the Goldsmith's Company. From around 1834 William II became a principal manufacturer for the Royal Goldsmiths Rundell Bridge and Co. He was joined by Daniel Ball at the famous workshops at 108 Bunhill Row off Chiswell Street in 1839.

13.11a JOHN BATSON & SONS 1880-1892

John Batson took up silversmithing around 1871. In 1880 he brought his two sons into partnership, which was dissolved in 1892 upon his retirement. The firm made a wide variety of objects including furniture, dressing cases and art metal work.

13.12   JAMES BEEBE 1811-1837

Following his apprenticeship to William Seaman he set up business making spoons at 30 Red Lion Street in Clerkenwell in 1811. In 1817 he moved to 26 Wilderness Row and then back to Red Lion Street but this time to number 65. He was a maker of good quality sauce labels.

13.13   BENT & TAGG 1832- 1850

The firm of Bent & Tagg entered its mark as silver platers in August 1832 in Birmingham. In 1834 the firm produced a fairly large label (4.7cm by 2.5cm) for DENTIFRICE WATER presumably to hang around a glass jar in the Dressing Room containing this toiletry.

13.14   RICHARD BINLEY 1745-1764

Having been apprenticed to Sandylands Drinkwater he became a specialist maker of labels, working in Gutter Lane. Unmarked crescent shaped OIL and VINEGAR labels have been attributed to him. Around 1760 he made and marked an ELDER label which can be attributed at the latest to the year 1764.

13.15   MARGARET BINLEY 1764 – 1778

Probably the widow of Richard Binley, she worked from 16 Gutter Lane, near Goldsmith's Hall in the City of London. She seems to have been a specialist maker of labels, passing over the business in 1778 to Susanna Barker. She was probably the earliest maker of Boudoir labels with romantic titles such as ALBA FLORA, AROMATIC VINEGAR, EAU DE COLOGNE, MILLE FLUERS and VERBENA. She made tiny long narrow rectangular labels with feather edged borders for boudoir relaxing refreshments such as the set of five for ALE, CYDER, PERRY, R CONSTANTIA and W CONSTANTIA. These labels are in the Marshall Collection, made around 1770 and are only a fifth or so of the size of the average wine label. As noted in "Sauce Labels" she made good use of technical developments mixed with a sense of style which makes her labels outstanding.

13.16   FREDERICK BRASTED 1862-1888

He was in partnership with John Bell from 1857 to 1862 at number 6 (renumbered 37 around 1880) President Street in Clerkenwell. He died in 1888. In 1875 he made a MILK PUNCH.

13.17   JOHN BRIDGE 1777-1834

Phillip Rundell took John Bridge into partnership in 1788. The firm became Rundell, Bridge and Rundell in 1805. John Bridge entered his own mark in 1823 (when Rundell retired) in four sizes. Whilst therefore he was a plateworker he nonetheless had a small punch for making small items, such as on a BRANDY label made around 1830.

13.18   BRITTON, GOULD & Co 1898 – 1910

This firm, whose premises were at 83 Hatton Garden, specialised in making novelties.

13.19   NICOLAI CHRISTENSEN 1820 – 1832

He was a Danish silversmith working in Copenhagen. He produced a fruit juice label for GUJAVA around 1830.

13.20   COCKS & BETTRIDGE 1773 – 1820

This firm worked from Church Street in Birmingham. In 1814 it produced a kidney-shaped label for GOOSEBERRY with a gadroon type edge.

13.21   JAMES COLLINS 1816 – 1838

James Collins entered his first mark in Birmingham

in October 1816 and his last mark in May 1838. It was his last mark which appeared on PEPPERMINT, an oval-shaped label with a rococo style foliage frame.

13.22   WILLIAM CONSTABLE 1806-1841
This Dundee silversmith along with David Manson followed in the tradition and standards set by Edward Livingstone in label making. He died in 1841 and his widow took over the tools of his trade. Dundee ranked with Edinburgh, Glasgow and Aberdeen as a producer and exporter of wine, sauce and boudoir labels. Constable made sauce labels for Anchovies, Ketchup and Soy as well as the boudoir label for CURRANT in 1807.

13.23   CRICHTON BROTHERS 1895 – 1956
For many years Crichton Brothers held Royal Warrants as Silversmiths, for example by appointment to King George V as Prince of Wales and subsequently to each reigning monarch. The firm specialised in antique reproductions which bear the LAC mark. The firm was established in 1895 at 29 Church Street, Kensington, moving to 22 Old Bond Steet. There were branches in Fifth Avenue, New York, and in South Michegan Avenue, Chicago.

13.24   WILLIAM CUNNINGHAM 1804-1828
This Edinburgh silversmith was a member of the large Cunningham family of silversmiths. William and Patrick Cunningham started around 1778. John, Alexander and Simon started around 1798-1800. Patrick, presumably the surviving partner, took his own sons into partnership in 1808. In the meantime William in 1804 and David in 1808 set up on their own account. Alexander's widow took over his workshop in 1812, and other Cunninghams formed partnerships with other silversmiths in Edinburgh.

13.25   CHARLES DALGLEISH 1816-1825
This Edinburgh maker, whose mark is known from some salt spoons, produced a FRONTIGNAN in 1817.

13.26   CHARLES HENRY DAVIS 1950-1970
This London silversmith made in 1960 the DANDELION label which is and was designed to be double-sided with RASPBERRY, thus making it a dual-purpose label.

13.27.   THOMAS DILLER 1828 – 1851
First entered his mark as a smallworker in 1828, working from 1 Richmond Building, Dean Street, Soho (Grimwade 2733). His second mark was entered in 1841 (Culme 13804). He was described as a Goldsmith when in 1829 his daughter Mary Ann was baptised at St Anne's Soho. He was in fact the founder of the famous firm of Harris & Brownett, goldsmiths and spring bottle cap makers in 1851. Thomas claimed to be the actual inventor of the spring-cap for scent bottles. Kitchen & Abud of 46 Conduit Street retailed around 1840 a gold-mounted ruby glass scent flask with a spring-cap which was stamped "Diller". A similar flask was retailed by Turners of 58 New Bond Street around 1845. Harris & Brownett made silver mounted scent flasks and silver-gilt mounted smelling salts flasks. Thomas Diller's HAIRWATER and EAU DE COLOGNE labels would have probably be hung on similar glass flasks. Other retailers included CF Hancock of Bruton Street, Howell Jones & Co of Regent Street and F. West of 1 St. James's Street who also supplied dressing cases which contained scent bottles.

13.28.   JOHN DOUGLAS 1804 – 1821
Starting off as a buckle maker in Clerkenwell (1788-1800) he entered into a partnership with Richard Lockwood in 1800 at 8 Clerkenwell Green. From January 1804 he was on his own until 1821, when he went into partnership with his son Archibald. This partnership made the silver fittings for a travelling case in 1825 and in later years (1848 – 1860) the mark of Frances Douglas has been noted on many silver mounts to bottles in dressing cases retailed, for example, by Charles Asprey of 166 New Bond Street. Portmanteau making was a concern of Archibald John Douglas (probably John's grandson) from 1864 – 1872 when this aspect was taken over by Bray & Willis. In 1809, John Douglas made a Campaign Box label for AURQUEBUSADE for use in the Napoleonic Wars.

13.29   JOHN DUDLEY 1846-1856
He entered in London two marks in 1846 and two further marks in 1847 as a gold workier, and was listed as a goldsmith in 1848. He worked from 2 Great Turner Street, Commercial Road East, just beyond the City boundary.

13.29a THOMAS EDWARDS 1816-1836
Thomas Edwards learnt the art of label-making by virtue of his apprenticeship to John Robins of Aldersgate Street in the City of London. He was then, it seems,employed as a journeyman for a long

time (1796 to 1816) before setting up on his own account as a plateworker in Islington.

13.30 WILLIAM ELLERBY 1802 – 1815

He was a smallworker working from 8 Ave Maria Lane, off Ludgate Street, in the City of London. He registered his first mark in 1802, another in 1804 but without any pellet and then a third mark in two sizes in 1810 within an oval punch.

13.31 WILLIAM ELLIOTT 1795 – 1809, 1813 – 1830

Thought to have been the son of silversmith William Elliott of Warwick Lane, he went into partnership with JW Story working at 25 Compton Street, Clerkenwell in 1809. Although the partnership lasted in 1815, Elliot entered two marks on his own account in 1813, used on a shell ANCHOVY label of that date and on a Campaign Box label for ARQUEBUSADE which bears makers mark only. In 1818 he overstamped another maker's mark on ARQUEBUSADE.

13.32 JOHN EMES 1796 – 1808

John Emes, after partnership with Henry Chawner 1796-8, entered his own mark and ran the firm until his death in 1808. He made a boudoir label for NECTARINE in 1799. In 1808 his widow Rebecca took the firm's manager Edward Barnard into partnership until 1828, after which her financial interest (and Henry Chawner's) continued, and the firm became Barnards.

13.33 WILLIAM WALKER EVANS 1872 – 1880

William Walker Evans was born in 1851, the year of the census, the son of William Evans (born 1819), who was a working jeweller operating from premises at 86 and 91 Bartholomew Close at West Smithfield and later on from the Cloth Fair nearby. His son took over in 1872 and moved to Aldersgate Street. The BITTERS label he made is dated 1876.

13.34 WILLIAM STEPHEN FERGUSON 1828 – 1875

Ferguson was apprenticed for seven years to Charles Fowler in Elgin in 1817. In 1823 he went to Edinburgh for two years (possibly spending a short time in Forres where he was born) to improve his silversmithing skills having just married. He then set up in Peterhead from 1825 until June 1828 when he moved to premises in High Street, Elgin. In the Elgin Courier he is described in 1828 as a goldsmith and watchmaker. It was said in the advertisement that

he would "always have on hand a choice assortment of Gold and Silver watches, Silver Spoons and every other article connected with the Jewellery business". He made good quality flatwear, including toddy ladles and sugar tongs, a snuff box and at least two boudoir labels for EAU DE PORTUGAL and VIOLETTE. According to the Merchant Guildry Minutes for 1833 he was "created and admitted goldsmith, jeweller and silversmith, admitted a Burgess and Freeman of the Burgh on 21st October 1833, and a Guild Brother of this Burgh for which on the 8th October 1828 he paid the then collector the sum of £4 sterling being the regulated dues of admission for apprentices of Guild Bretheren". He became an influential member of the Burgh and held various public offices.

13.35 RICHARD FERRIS 1784-1812

This Exeter silversmith who had been apprenticed to Thomas Eustace, flatware specialist, made a wide range of practical tableware, mainly spoons and small domestic items, so it is not surprising to find him making silver labels for FRUIT and GINGER. His main production was during the period 1795 to 1808, during which time he made many labels, often attractively engraved.

13.36 GEORGE FISHER 1866-1880

George Fisher started in 1866 in Clerkenwell at 54 Rahere Street before moving in 1873 to City Road at 122 Central Street. His mark appears upon a VINEGAR label of 1877.

13.37 WILLIAM FOUNTAIN 1794 – 1823

He and his elder brother, John, were apprenticed to Daniel Smith of Aldermanbury in the City of London. Having obtained his freedom in 1785, he worked as a journeyman plateworker in Charterhouse Lane until in 1791 he went into partnership with Daniel Pontifex in West Smithfield. In 1794 he went solo at 47 Red Lion Street, Clerkenwell. Towards the end of his career he made a splendid looking Hartshorn boudoir label.

13.38 THOMAS HENRY FRANCIS and FREDERICK FRANCIS 1854-1866

These two brothers were the sons of Fleet Street silversmith Thomas Francis. Thomas Henry was born in 1825 and Frederick in 1829. They occupied manufacturing premises in the Oxford Street area from 1852 to 1854 and then moved to 24 Poland

Street, Soho. Their business was however closed down due to debt in 1866. However they made important dining room related silver. One of their pieces was exhibited in the 1862 International Exhibition in Class 33 (numbered 6622).

13.39   CRISPIN FULLER 1792-1823
He was a plateworker in Monkwell Street, starting off at number 42 in 1792 and ending up at number 3 Windsor Court which leads off Monkwell Street.

13.40   JOSEPH GIBSON 1784-1820
He registered two marks with his surname contained within a rectangular punch numbered 27 and 28 by Douglas Bennett. He was described as a goldsmith, jeweller and watchmaker. He made wine laels and a rather splendid pair of nips.

13.41   JOSEPH GLENNY 1792 – 1821
He used incuse marks, perhaps because he also made watches. He spent eight years at 13 Old Street Square before moving to 22 Charles Square, Hoxton. Then, like so many other silversmiths, he ended up in 1804 in Clerkenwell. He liked floral designs and incorporated exotic fruits including pineapples in his artistry. He was also a sauce label maker, particularly known for his ANCHOVY of 1819.

13.42   SAMUEL GODBEHERE, EDWARD WIGAN, JAMES BULT 1800-1818
S. Godbehere & Co was entered in the register as S. Goodbehere & Co. on 15th March 1800, working from 86 Cheapside in the City of London. Samuel signed as Goodbehere. James Bult is also recorded as James Boult, to add to the confusion. The firm had connections with Bath and Bristol, Edward Wigan's father being a Bristol goldsmith, and especially with William Bottle and James Burden.

13.43   ROBERT GRAY 1776-1806
The earliest date of mention of Robert Gray as a silversmith is in 1776. About this time he made a soup-ladle, a hash-spoon and a teapot. Around 1807 he brought his son into partnership. The firm made tea-spoons, wine-labels, castors, badges, lemon strainers, beer jugs, spirit lamps, mugs, sugar-tongs and such like. In the two decades or so before the establishment of the Glasgow Assay Office in 1819 his silver was assayed in Edinburgh. This included a CORDIAL label of around 1805. The Glasgow Goldsmiths' Company commissioned Robert

Gray & Son to make a two-handled cup and cover for presentation to Mr. Kirkman Finlay, the Lord Provost, who was MP for Glasgow Burghs from 1812 to 1818 and who helped to get Parliamentary support for the Private Act which set up the Glasgow Assay Office.

13.44   ARMAND GROSS 1893 -1899
Armand Gross entered his mark in Paris on 31st October 1893. He chose a diamond shaped punch with his name in it around a winged wheel. He took over a shop at 158 Rue Du Faubourg – Saint – Martin from M. Ferry in 1893.

13.45.   THOMAS HALFORD 1807 – 1832
Thomas Halford was the son of a haberdasher and was apprenticed to a goldsmith, Alexander Field of Hoxton. It seems that from 1786 to 1807 he worked as a journeyman plateworker. He opened his own business in 1807 at 34 City Road and then in 1829 he moved to Stratford.

13.46   HAMILTON & INCHES 1866-2012
The firm was founded in 1866. Their mark was registered in Edinburgh in 1880, and a version in script appeared in 1899. They were prolific makers of quality items including a baton for the Duke of Argyll in his capacity as Master of the Household. They are Royal Warrant holders and have premises in London.

13.37   HAYNE & CO 1836 – 1848
Founded by Thomas Wallis Senior around 1758, Hayne & Co as this major firm of manufacturing silversmiths was styled between 1836 and 1848 was run by Samuel Holditch, John Hayne and Dudley Frank Cater, the latter being a former apprentice, but Jonathan Hayne (Prime Warden of the Goldsmiths' Company in 1843) continued to maintain a substantial share in the business, as shown by the terms of his will dated 11th December 1841, until his death in 1848, after which the firm was known as Hayne & Cater until 1864. Hayne and Cater produced a label for CAMPHORATED SPIRITS OF WINE in 1843.

13.48   THOMAS HAZLEWOOD 1888 – 1915
Thomas H. Hazlewood entered his first mark in Birmingham in January 1888, his second mark in May 1907 and his third mark in May 1912. T H Hazlewood &Co was formed in 1892 and entered marks in Birmingham in 1892, 1896, 1899, 1906 and

1911. The boudoir label for METHYLATED bears the mark of Thomas himself and not the company with a date letter for 1913.

13.49   DANIEL HOCKLEY 1810-1819
He was born in London in 1788, the son of Thomas Hockley of Seven Dials, an oilman. He was apprenticed to John Reily in 1801 and became a Freeman of the Fishmongers' Company in 1808. His workshop was at 9 Brook Street in Holborn, where he made labels of quality. Having decided to emigrate to South Africa it seems that he sold his business to Charles Rawlings in 1819. He died in Graaf-Reinet in 1835.

13.50   JOHN HOLLOWAY 1772-1795
He was a smallworker based at 24 Witch Street, Clement Danes. It seems that he was apprenticed to John Wirgman in December 1754. He did not however gain his freedom until March 1770. He is the attributed maker of a label for S. WATER, whatever that stands for – Spring, Soda, Seltzer are some of the possibilities.

13.51   SOLOMON HOUGHAM 1788 – 1790, 1793 – 1817
He was a goldsmith and was elected a Liveryman of the Goldsmith's Company in 1791. He was in partnership with his brother Charles from 1790 until his death in 1793. his workshop was in Aldersgate Street in the City of London. In 1817 he went into partnership with Solomon Royes and John East Dix.

13.52   WILLIAM HOWDEN 1801 – 1809, 1824 – 1828
Thought to have been a relative of Francis Howden (1781-1801), a label maker who appeared in the Edinburgh Directory of 1795 as working from 3 Hunter's Square, William worked in Edinburgh from 1801 to 1809, when he joined up with James Howden in partnership. In 1814 a firm was created called J. W. Howden & Co. William then registered on his own account in 1824. Then in 1828 a re-organisation took place. A new firm known as James Howden & Co was formed which remained active until at least 1855. William's GINGER label is attributed by Strafford and Thomas to around the year 1805.

13.53   HUKIN & HEATH 1881 – 1953
This firm of manufacturing silversmiths and electroplaters was established in Birmingham in 1855 at Imperial Works 137-139 Great Charles Street. It opened London showrooms at 19 Charterhouse Street in 1879. Hukin retired in 1881. The firm was continued by John Thomas Heath and John Hartshorne Middleton using a joint punch until incorporation was achieved in 1904 with limited liability. They made an ASTRINGENT in 1908. Middleton seems to have retired or perhaps died before 1909. Heath, a former Chairman of the London Wholesale Jewellers' and Allied Trades' Association, died in December 1910. The firm gave great encouragement to young designers, with the help and assistance during the period 1887 – 1890 of Dr. Christopher Dresser.

13.54   THOMAS HYDE I 1747-1804
Thomas Hyde's father James was a vintner who had many contacts and clients requiring wine labels. His specialisation can be traced back to the Binleys and Sandylands Drinkwater. He was apprenticed to John Harvey, whose premises at 33 Gutter Lane he acquired in 1747. He had two sons, Thomas and James. Thomas used his father's punches! So the VINEGAR label of around 1775 could have been the work of father or son, but has been attributed to Thomas Hyde I.

13.55   JAMES HYDE 1777 – 1799
He was a well known and respected maker of labels, working from 10 Gutter Lane and 38 Gutter lane in the City of London. Number 38 was directly opposite number 10. He then worked at 6 Carey Lane nearby from 1796 until his death in 1799.

13.56   ALEXANDER JOHNSTON II 1760 – 1799
This somewhat shadowy figure is recorded as being made free by redemption on 9th February 1757. He was described as a goldsmith. Heal records him as being with Charles Geddes jewellers of the Golden Ball in Panton Street near Leicester Square around 1760. He is however referred to as a plateworker in the Parliamentary report list of 1773.

13.57   CHARLES KAY 1815 – 1827
Charles Kay made silver, including a CAYENNE sauce label, from premises in Addle Street, Aldermanbury, from 1815 to 1819, in New Street, Blackfriars, from 1819 to 1823, at 11 Addle Hill from 1823 to 1825, at 12 Pump Row, Old Street, from 1825 to 1827 and finally at 14 John's Row from 1827. Such mobility is unusual, even for a small worker.

13.58   DARBY KEHOE 1771-1780
He is the attributed maker of a FRONTIGNAC label in Dublin. He was made free in 1771 as a goldsmith and in that very same year took as an apprentice Robert Cubbin from the Isle of Man. His mark appears on some tablespoons of 1777.

13.59   GEORGE KNIGHT 1818 – 1825
George Knight's mark, entered in 1818, is similar in form to that of Samuel Knight entered in 1816, from which it may be deduced that there was some form of relationship. They both worked from Westmoreland Buildings in Aldersgate Street. His ORANGE of 1820 is slightly curved, heavy, with a triple reeded border and pierced lettering.

13.60   SAMUEL KNIGHT 1810-1827
Like George Knight he was a label maker working in Aldersgate Street which was within the City of London and from which it was easy to supply the shops. He made a label for GRAPE in 1816.

13.61   WILLIAM KNIGHT 1810-1846
George, Samuel and William Knight were in some way related, probably brothers. Samuel seems to have been in partnership with William in 1810. George's mark entered in 1818 in similar form to that of Samuel entered in 1816 and they both for a time worked in 7 Westmoreland Buildings in Aldersgate Street. In 1816 William registered his own mark, working from Bartholomew Close until 1827 and then from 7 Westmoreland Buildings until 1846.

13.62   LAMBERT & Co 1861 – 1916
Francis Lambert founded the firm of Lambert & Co. He started working from 1803 at 11-12 Coventry Street. William Rawlings joined him in 1820. The firm was created in 1861. Towards the end of the Nineteenth Century the firm was led by Colonel George Lambert FSA and established a reputation as a firm of jewellers and retail silversmiths. Herbert Charles Lambert (1902-1912), who was George's nephew, produced two silver-gilt enclosed crescents in 1908 perhaps for use on an elegant oil and vinegar frame but possibly, being of silver-gilt, for use on small glass bottles contained in a travelling box which would bring them into the boudoir category. The business closed in 1916.

13.63   LEA & Co 1811 – 1824
Lea & Co registered its mark in Birmingham in November 1811. In 1821 it made a label for COWSLIP. In June 1824 a new firm was formed following the merger with Clark known as Lea and Clark.

13.64   LEVI & SALAMAN 1870 – 1921
Levi & Salaman, founded c. 1870, were a leading firm of Chasers and Stampers based in Birmingham. They registered several marks. In 1921 Barker Brothers, a leading manufacturer of EPNS, and the Potosi silver company, a firm of electroplaters, merged with Levi & Salaman in order to cope with the post World War I economic depression.

13.65   HERMANUS LINTVELD 1817 – 1833
This Dutch silversmith employed the rare shield shape design for ANISETTE and also for CURACAU. He may well be the maker of the unmarked set of the rare shield shape design labels for BRANDY, SHERRY and HOLLANDS which may have been used in the boudoir.

13.66   MATTHEW LINWOOD II 1793 – 1821
He was the eldest son of Matthew Linwood I (1773-1783) and an active label maker in Birmingham. In 1793 he promoted the die-struck process to create mass production of labels. He entered his mark in Sheffield in 1808 to develop his interest in plate making. In 1811 he served as Guardian of the Assay Office in Birmingham. The year afterwards he produced an ARBAFLOR perfume label.

13.67.   EDWARD LIVINGSTONE 1792-1810
Edward Livingstone worked in Dundee in Scotland producing mainly flatware in competition with James Douglas. Around 1795 he produced the small rare star-shaped ACID.

13.68   GEORGE LOWE 1791 – 1841
The Chester Plate Duty Books record that in 1839 George Lowe submitted ten labels and ten bottle labels for assay. Previous descriptions in the Duty Books refer to bottle tickets, labels and wine labels. His family made a considerable contribution to silversmithing in Chester and he was appointed Assay Master in 1840.

13.69   DAVID MANSON 1809 – 1820
David Manson was admitted in Dundee in Scotland in 1809. He made mainly flatware as well as a Boudoir label for GINGER.

13.70   MAPPIN AND WEBB 1859 to date

Mappin & Webb 1913 Limited is a well known firm of manufacturing and retail silversmiths. The original firm was established by John Newton Mappin in partnership with George Webb in 1859. In 1886 the firm opened workshops for the manufacture of travelling bags and dressing cases at numbers 1 and 2 Winsley Street off Oxford Street known as the Winsley Works. A journalistic report on a visit to its factory in Sheffield in 1886 notes that on display there were "a fine assortment of dressing-cases, made at the London warehouse, and fitted with requisites produced by the firm in Sheffield ….. and various novel appliances, such as holders for champagne bottles, with handles after the fashion of a claret jug, and trays fitted with all the requirements for toddy". In 1908 the firm filled a stand in the Palace of Woman's Work at the Franco-British Exhibition with articles for use by women which included toiletries. In 1914 they produced a boudoir label for METHYLATED SPIRIT.

13.71   HENRY MATTHEWS 1874-1897

This Birmingham silversmith, with an elegant maker's mark which gives individual treatment to each initial, is the maker of a set of three labels for ANNISETTE, CHARTREUSE and COGNAC, made in 1880.

13.72   NATHANIEL MILLS 1825-1848

Nathaniel Mills of Caroline Street Birmingham was one of the finest and most prolific boxmakers of his time. Examples of his work can be seen in the Birmingham Assay Office, including vinaigrettes, card cases, memorandum cases, snuff boxes and sweetmeat baskets. Public taste for souvenirs tended to dominate some of his designs. He made a BRANDY label for use in the boudoir in 1845.

13.73   WILLIAM MOODY 1756-1786

He had the privilege of being apprenticed to Charles Frederick Kandler of Jermyn Street, St. James, in 1743. William Moody entered a mark as a largeworker in 1756 giving an address in Berwick Street, St. James. He was still there in 1773, recorded as a smallworker in the Parliamentary List of that year. He is recorded as being in Taylor's Street in 1758 and possibly in Aldersgate Sreet in 1786. He made a label for CURRANT.

13.74   ELIZABETH MORLEY 1794 – 1814

A widow, she was prolific label maker, operating from 7 Westmoreland Buildings. She used at least twelve different styles of label design and decoration.

13.75   JAMES NEWLANDS and PHILIP GRIERSON 1811-1816

James Newlands entered his first mark in 1808 and went on silversmithing until some time after 1819. He was the sixth signatory (signing was in order of precedence) to the 1818 Petition to parliament for the Incorporation of Glasgow Silversmiths. Philip Grierson entered his mark in 1810 and joined James in partnership in 1811. However in 1816 they appear to have split up and gone their separate ways.

13.76   GEORGE PEARSON 1817-1825

George Pearson was a little known smallworker in Fleet Street working from 104 Dorset Court. His only recorded mark was entered in 1817 as recorded by Grimwade (number 871). About this time he made a label for BRANDY.

13.77   B. PETERSEN & Co. 1900

According to Shepherd's Gold and Silversmiths in New Zealand silversmithing in that country dates back to the Nineteenth Century. The first craft exhibition was held in Dunedin (South Island) in 1865. There is and was no system of marking laid down by law. Petersen himself was based in Christchurch, South Island. He made ANGELICA and C. PORT.

13.78   S   P   1987

This London silversmith made the good looking MAPLE SYRUP label with the cut-out stag in 1987.

13.79   PHIPPS & ROBINSON 1783 – 1811

The famous partnership of Thomas Phipps and Edward Robinson both of whom had been apprenticed to and learnt from James Phipps was prolific in the manufacture of labels. The partners entered their first mark in 1783, the year of Edward Robinson's marriage to Ann Phipps. They operated from 40 Gutter Lane near Goldsmiths Hall in the City of London.

13.80   PHIPPS, ROBINSON AND PHIPPS 1811-1816

In 1811 they produced the interesting BRISTOL WATER label.

13.81   THOMAS & JAMES PHIPPS 1816 – 1823
This effective partnership carried on the Phipps and Robinson tradition. The partnership came to an end with the death of Thomas Phipps in 1823. They made an ESPRIT DE ROSE in 1820.

13.82   JOSEPH PRICE 1821 – 1838
The son of a poulterer, he gained his Freedom in 1815. He went into partnership with his brother William in January 1820 as a smallworker. The partnership lasted until August 1821. He stayed on at 5 Silver Street, Wood Street, until 1825 when he moved to 49 Red Lion Street, Clerkenwell.

13.83   CHARLES RAWLINGS 1817 – 1829
Charles Rawlings was apprenticed to Edward Coleman, a watch finisher. He was a prolific maker of labels operating from the former premises of James Atkins and later his widow Theodosia Ann Atkins at 12 Well Street, London. In 1819 he probably took over Daniel Hockley's business at 9 Brook Street. Charles is said to have introduced the famous vine leaf design for wine labels around 1824. He was certainly innovative.

13.84   RAWLINGS & SUMMERS 1829 – 1897
Charles Rawlings had moved to 9 Brook Street Holborn in 1819 and it is from this address that he began his long partnership with William Summers. They were so successful that they moved to the west end of London and opened premises near Regent Street at 10 Great Marlborough Street. Charles died on 9th October 1863 and William on 15th January 1890. Their great contribution to labelling is illustrated by the following examples: ANISEED 1829, COGNAC c1830, EAU DE ROSE 1830 (SG), ELDR. FLOR. WATER 1830 (SG), ESS. GINGER c1830, EAU DE COLOGNE 1831, EAU DE PORTUGAL 1831, LAVENDER WATER 1831 (SG), ESCHALOTTE 1832, EAU DE PORTUGAL 1834, ROSE WATER 1836, LAVENDER WATER 1836 (SG), FLEUR D'ORANGE 1837 (SG), PARFAIT AMOUR 1837 (SG), DISTILLED WATER 1838, GRAPE 1839, B (for BLACK INK) 1840, WATER c1843, ARISETTE c1844, ANISETTE c1845 and NOYEAU 1856.

13.85   JOHN REILY 1801 – 1826
John Reily was the son of Richard Reily, a glazier in Thames Street. He was apprenticed to James Hyde in 1786 and admitted a Freeman of the Fismongers

Company on 13th February 1794. John Reily first entered his mark in partnership with James Hyde's widow in 1799 and then on his own account. He married Mary Hyde in 1801. He took on Daniel Hockley and George Pearson as apprentices.

13.86   MARY & CHARLES REILY 1826-1829
John Reily's widow Mary went into partnership with their son Charles in 1826, the partnership only lasted three years and in 1826 it produced a boudoir label for NOYEAU.

13.87   REILY & STORER 1828 – 1855
Charles Reily around 1828 entered into partnership with George Storer who had in fact served his apprenticeship as a watch maker. This partnership made a wide variety of boudoir labels, some with maker's mark only as in the case of CAMPH DROPS, some without marks and thus attributed as in the case of EAU D'ALIBURGH.

13.88   JOHN RICH 1765-1810
John Rich was a prolific maker of labels at his workshop near Whitfield's chapel in Tottenham Court Road, so much so that although he died in 1807 it took three years to dispose of all his stock.

13.89   RICHARD RICHARDSON IV 1779-1822
It has been said that Richard Richardson IV may have been the maker of a set of three unmarked labels for FRONTINIAC, MADEIRA and RAISIN. They are not recorded by Ridgway. Two members of the Richardson family did indeed make labels, mainly for Port around 1767, 1781 and 1790. The attribution seems unlikely.

13.90   ROBB & WHITTET 1855 – 1880
The dates shown are very approximate. Their mark is known on a piece of plate in Goldsmiths' Hall in Edinburgh dated 1876. The label for SWEET has only the duty paid mark of Queen Victoria and the thistle mark with R & W in a shaped rectangular punch.

13. 91   WILLIAM ROBERTSON 1795-1805
This Edinburgh silversmith made a label for SOURING around the turn of the Eighteenth Century.

13.92   JOHN ROBINS 1774-1801
He entered a mark (Grimwade 1623) in 1774. He operated from 67 Aldersgate Street from 1781 to

1794. He then moved to 13 Clerkenwell Green.

13.93   PAOLO ROSSO 1857 – 1870

Paolo Rosso received his warrant to work as a silversmith in Malta on 14th October 1857. Unusually, he registered two marks, one for PR in the common rectangular punch and one elegant scripted initials within an oval punch which he used on BITTERS, made around 1860. On the 27th October 1862 he registered a third mark, the elegant scripted initials again but this time the oval was tuned onto its side. The BITTERS label bears the Maltese Cross mark introduced on the 1st of January 1857 for the standard of silver at 11 deniers reduced from 11.5 deniers.

13.94   R . W. SMITH 1818-1850

R.W. Smith, whose forenames do not seem to have been recorded, was made free of the Dublin Silversmiths' Company in 1818. His mark is recorded on a copper plate covering the period 1813 to 1850. It is very distinctive, rather like a rectangular shaped flag. It is possible that in the year before it was registered he was in partnership with W. Hamy.

13.95   JOSIAH SNATT 1797-1817

He worked from 4 Fann Street, Aldersgate, in the City of London making good quality labels, including one for SAL VOLATILE in 1801.

13.96   ALEXANDER STEWART 1820 – 1830

Alexander Stewart was a Highland silversmith selling wares in Cromarty, Inverness and Tain amongst other burghs.

13.97   STOKES & IRELAND 1878-1935

Stokes and Ireland entered their first mark in Birmingham in September 1878. The firm was incorporated as a Limited Liability Company, entering its mark in July 1892. The company merged with S. Blanckensee & Son around 1935 having suffered during the time of the great depression.

13.98   PAUL STORR 1792-1838

This famous maker of fine display plate was apprenticed c.1785 to Andrew Fogelberg. He thus began his career in Church Street, Soho, in 1792. He moved out of the Soho area to Harrison Street off the Gray's Inn Road in 1819, having severed his connections with Rundell, Bridge and Rundell. His last mark of many was entered in 1834. He retired to Tooting in 1838.

13.99   THOMAS STREETIN 1794 – 1830

Thomas Streetin was a plateworker working in Camberwell Green. He was the son of a watchmaker and apprenticed to a spoonmaker. In 1816 he took his son Henry as an apprentice. In 1829 he made a pair of interesting boudoir labels for use in the travelling case.

13.100 WILLIAM SUMMERS 1859 – 1887

After the retirement of Charles Rawlings in 1859 (he died in 1863) Summers carried on the traditions of Rawlings and Summers, reviving in Victorian times styles popular in earlier periods. At the beginning of his sole responsibility he produced VIOLET in 1859 and at retirement in 1887 (he died in 1890) he produced EAU DE COLOGNE and LAVENDER WATER. His firm was a major influence in promoting continued interest in the use of silver boudoir labels.

13.101 JOSEPH TAYLOR 1780-1827

Joseph Taylor of Birmingham has been described by Kenneth Crisp Jones as a fairly typical all-round silversmith. His Trade Card of 1816 lists over 60 different objects. He entered his mark in 1780 and died in 1827, working in Newhall Street as a goldsmith, silversmith and jeweller. He made a variety of small objects including scent boxes and smelling bottles. The Wellcome Institute of History of Medicine has a silver scent bottle case made by him in 1807. In 1821 he made a pair of interesting broadly oval shape stamped labels for CLARET and MADEIRA with scroll and vine motifs with a satyr mask below the engraved title plaque.

13.102 HENRY THORNTON   1885-1912

Henry Thornton entered his mark in London in 1885, giving as his address 48 Walton Street, Chelsea, where a Henry Thomas Thornton, presumably his son, was listed as a watchmaker in 1897 and 1913. Henry Thornton made some wine labels, an A1 SAUCE label in 1895 and a set of labels for BENEDICTINE, MARASCHINO, CHERRY BRANDY and GINGER.

13.103 CHARLES TIFFANY 1891-1902

Tiffany & Co. was incorporated in 1868. Its success

was largely due to its founder President Charles Louis Tiffany and its silver designer Edward C. Moore with whose firm a merger took place. Its extensive workshops were at 53-55 Prince Street New York. A beautiful boudoir label for GUM SYRUP bears the Tiffany mark used between 1891, the death of Edward C. Moore, and 1902, the death of Charles L. Tiffany, when the President was Charles L. Tiffany indicated by the impressed capital letter T being the first letter of his surname. The Company made interesting labels. 12 Wine Labels were included in Chest1 of the Mackay Service containing a formidable array of 918 pieces of flatware. Silver Wine Labels were offered for sale in the "Presents for Gentlemen" section of the 1880 Blue Book at prices ranging from $1.50 to $15. Glass toilet bottles for men and women and "Odor Bottles" with richly cut glass for women in silver mounts specified in the 1880 Blue Book afforded a great opportunity for labelling. A splendid Tiffany CHARTREUSE has been illustrated in the Journal (1 WLCJ 5, p 72).

13. 104 JOHN TOLEKEN 1795-1836
This silversmith worked in Cork and tended to mark his products with a TOLEKEN punch accompanied by his initials or the word STERLING. He is recorded as having made a sugar bowl as well as wine labels around the year 1800 in Cork. Unfortunately the Cork Guild and Goldsmiths' Company records were all destroyed by a fire which occurred at the Cork Courthouse on the 27th march 1891.

13.105 WILLIAM BAMFORTH TROBY 1804-1829
William was the son of John Troby, a smallworker from Ship Court near the Old Bailey in the City of London. He became a plateworker, but none-the-less made sauce labels (one with an ambassadorial connection) and a boudoir label for CORDIAL, a most important refreshment.

13.106 WILLIAM TWEMLOW 1790 – 1805
William Twemlow made a wide variety of silverware in Chester. In 1799 he produced a label for FRUIT and a label for GINGER.

13.107 GEORGE UNITE 1824 – 1928
A prolific label maker, George Unite worked in Birmingham. He was apprenticed to Joseph Willmore and then to James Hilliard with whom he entered into partnership in 1825. This ended in 1832 and he was on his own. In 1865 he founded the firm known as George Unite & Sons. It prospered, supplying silver for example to Hamilton & Co in India in the 1880s and making a wide variety of flatware and small items. The firm was incorporated in 1910 and merged with Lyde & Co in 1928 when a new Limited Liability Company was formed.

13.108 UNKNOWN MAKER 1900
A pair of boudoir labels for the perfumes VIOLETTE DE PARME and WHITE ROSE were hallmarked in Birmingham but the makers mark was not recorded.

13.109 WAKELY and WHEELER 1930 – 1960
This firm, based in London and Birmingham , is well known for its Ascot Gold Vase of 1939 which was designed by R.G. Baxendale. So it is not surprising that its set of five silver-gilt boudoir labels, made in 1956, is of fine quality. In 1953 it was commissioned by the Chester Goldsmiths' Company to make a silver-gilt half-pint tankard to Britannia standard to mark the Coronation of HM Queen Elizabeth II.

13. 110 THOMAS WATSON 1801-1811
Thomas Watson & Co of Sheffield hallmarked candlesticks in 1801 and made a label for RASBERRY in 1806.

13.111 JOSEPH WILLMORE 1806-1845
This famous Birmingham maker produced a wide range of silver objects and was prolific in label production. In 1845 he incorporated his business under the title of Joseph Willmore & Co. From 1830 to 1840 he made outstanding use of natural motifs as illustrated by plates 15, 16 and 18 of Kenneth Crisp Jones' "The Silversmiths of Birmingham". Label making was a challenge for him in the earlier part of his career, learning from Thomas Willmore and the Linwoods.

13.112 YAPP & WOODWARD 1845-1874
Both John Yapp and John Woodward entered into partnership with Joseph Willmore (see above) in 1834. However when the business was incorporated in 1845 Yapp and Woodward left and formed their own partnership.

Fig 14.1

CHAPTER 14

DESIGNS

The subject of dating by design is fully dealt with in Chapter 7 of "Sauce Labels". Therefore the following commentary is based on the principles set out in that Chapter. The designs have to fit the variety of bottles used, as shown in fig. 14.1 showing the water carafe, the medicine bottle, the perfume bottle and the cordial flask.

14.1   THE EYE

The eye, pointed oval or navette shape is an early design. When, as in the example of POISON, it is accompanied by double reeding as a border decoration, the period of production is approximately confined to the years 1792 to 1806. See figs. 9.217 and 9.226.

14.2   THE OVAL

The oval shape appears to have been used mainly between 1775 and 1860, a very wide bracket, and there has been a modern revival. This shape was chosen for a large porcelain WHITE ROSE. Many of the titles noted are in the French language which is indicative of a French origin. There exists, for example, a pair of ovals entitled HUILE DE VENUS and EAU DE COLODON. A set of eight ovals bears the unusual names of AQUA D'ORO, CRÈME DE CFE, CRÈME DE NOYEAUX ROUGE, CITRONELLE, ELIXIR, ELIXIR DE SPA, GRANDE MAISON and HUILLE DE VANILLE. More modern large ovals for Cloroform in blue script, GIGGLEWATER in blue capitals and PEPPERMINT and CLOVES in plate add interest. See fig. 9.56.

14.3   THE COLLAR

The collar shape or hoop, which fits around the neck of a bottle (hence "neck-ring"), did not come into production before 1790. A set of eight splayed hoop silver neck ring collar labels (see 1 WLCJ, p 278) fitted snugly over over a set of eight perfume bottles in a shagreen – or shark's skin – cased perfume travelling box made around 1800. The nature of the contents of each bottle could be easily viewed from above, giving an instant appreciation of ATTAR OF ROSES, BERGAMOT, CITRON, CLOVES, LAVENDAR, MINT, ROSEMARY and WOODBINE. Another splayed hoop silver label, entitled EAU de COLOGNE, is in the Cropper Collection held by the Victoria and Albert Museum. The title itself is picked out in niello. The hoop bears its owner's initials "ELB". Collars were also made of ivory, a material which lent itself to easy production of hoops, and an example exists for GINGER. Thomas Diller, wishing to be stylish, invented a new type of neck-ring for an EAU DE COLOGNE bottle. See figs. 9.84a, 9.86a and 9.126b.

14.4   THE BROAD RECTANGULAR

An unmarked ROSE in the style of James Beebe has an unusual raised foliate border. Compare, for example, his HARVEY illustrated in Sauce Labels in Photograph 38. The shape was, it seems, only used between 1800 and 1830, a narrow band. The style of the heavy gadrooning suggests that alternatively the maker could have been Daniel Hockley (1810-1817). The fact that ROSE is in silver-gilt and has pierced lettering puts it into a production period of 1809 to 1830. Labels with foliate borders look good in this design, see for example CORDIAL and GARUS. See figs 9.49, 9.136, 9.194, 9.210.

14.5   THE OCTAGONAL

This is the rectangular shape with cut corners. See figs. 9.15 and 9.22.

14.6   THE ROUNDED END RECTANGULAR

This common shape, often embellished with a double reeded border, was popular during the years 1790 to 1830. Thus the unmarked CAMPHOR JULEP, slightly smaller at 2.5cm by 1.3cm than Reily and Storer's similar labels at 2.6cm by 1.4cm, is datable to around 1790-1830. See figs. 9.11 and 9.45.

14.7   THE LONG NARROW RECTANGULAR

Examples can be cited showing off this design. In

silver there is Crème De Noyeau and Crème De Thé, in porcelain there is EAU FROIDE and in silver plate there is MELOMEL. See figs. 9.69 and 9.102.

14.8 THE LEAF
Rawlings and Summers used a vine leaf lying on its side in 1831 for EAU DE COLOGNE, EAU DE PORTUGAL and LAVENDER WATER. The vine leaf design was also used for WATER and for Rose. Whilst the vine leaf shape was thus used for wine, sauce and boudoir labels, and the oak tree leaf was used as a regimental wine label, the perfumier of the 1830s chose the oak leaf to adorn a set of three boudoir labels to adorn scent bottles containing ESS. BOUQUET, LAVENDER and OPPOPONAX. The design is very art deco in vogue from the 1920s until the privations of the Second World War and shortage of supplies had their effect. See figs. 9.86 and 9.218.

14.9 THE STAR
Edward Livingstone's ACID follows this design. Another version, also for ACID, made of mother-of-pearl in crescent shape, has a border of 12 six-pointed stars. See fig. 9.1 and "Wine Labels" figs. 231-233.

14.10 THE CUSHION
Margaret Binley's ALBA-FLORA is of the early humped cushion design, embellished with a feather edge. See fig. 9.4.

14.11 THE CRESCENT
The peaked crescent was a popular shape for enamels. The peak itself takes on a variety of interesting shapes. So Huile de Rose has a rather unusual kind of peak. In silver the design was adopted by Robert Gray in making his CORDIAL label. Solomon Hougham used a rounded end crescent infilled by an oval cartouche for his ARQUEBUZARDE. William Summers reproduced the crescent design for his COGNAC. The mother-of-pearl GRAPE is somewhat reminiscent of a sauce label crescent design. Wide open crescents were also used, such as those for MALVERN WATER, TOILET WATER and NEW MOWN HAY. See figs. 9.120, 9.134, 9.144 and 9.169.

14.12 THE SHIELD
This shape was quite popular for boudoir labels although rare amongst English wine labels but more popular on the Continent of Europe, such as for the BRANDY, HOLLANDS and SHERRY, unmarked but attributed to Hermanus Lintveld of around 1825. So this was the design chosen by Lintveld for his ANISETTE. See fig. 17.1 for German shields. It was chosen for the unmarked VANILLE. The Vintners' Company has the distinctive elongated version for CASTOR OIL. A triangular shield shape has been used for LAVANDE and a breast-plate shape for TONIC and VANILE. Chartreuse has an extended shield shape. The shield design therefore comes in a variety of shapes. See figs. 9.51, 9.62, 9.152a, 9.204, 9.215 and 9.237.

14.13 THE HEART
The romantic heart shape lends itself to boudoir labels. The most evocative is perhaps Levi and Salaman's version in silver for Eyes produced in 1901 in art nouveaux style. Unmarked French porcelains for EAU DE COLOGNE and LAVENDER WATER could have been a suitable Valentine's present. A Chinese style heart shape was used for ROSE WATER (Bonhams sale of the Lank Collection 5.10.2010) as well as for KUMMEL, CHARTREUSE, SHERRY and HOCK, suggesting that these were popular boudoir tipples. Three roses adorn the title which is then set in a sort of shield. See figs. 9.86f, 9. 113 and 9.219e.

14.14 THE PLAQUE
This design involves elegant scrolling or foliate baroque as used by William Evans for his BITTERS, Rawlings and Summers for their ARISETTE, George Lowe for his EAU DE COLOGNE, Thomas and James Phipps for their ESPRIT DE ROSE, and Robb and Whittet for their SWEET. It was also used for the plated LACHRYMA and LOVAGE. No less a man than Paul Storr used the rococo scroll cartouche in silver-gilt in 1835-1836 in small size for boudoir and sauce labels and in large size for alcoholic beverages, cast from their respective moulds. See figs. 9.13, 9.16 and 9.32.

14.15 THE KIDNEY
This shape beloved by Margaret Binley was quite popular. Heath and Middleton used it for their broad version of ASTRINGENT. Hukin and Heath in Birmingham in 1936 used it for their Peach in script lettering. The shape was also employed by enamellists and examples are labels for LAVANDE, QUETSCHE and FINE NAPOLEON. See fig. 9.20.

14.16   THE RIBBON

This design for enamels and porcelains was employed by Samson in two sizes of enamels – see MON PARFUM, WHITE ROSE, CRÈME DE VANILLE, DACTILUS, EAU BORIGUEE, EAU DE LUBIN, JASMIN, VIOLET, MOSS ROSE, WHITE LILAC and PORT, SHERRY and CLARET, when drunk in the boudoir as a favourite tipple. It is also featured on Staffordshire and Coalport porcelains – see WHITE ROSE and METHYLATED SPIRITS. See fig. 9.72.

14.17   THE SLOT

Made popular by Messrs Floris, slots were available in silver or pewter and had ivory or parchment titles. The advantage given by a slot was that it was relatively easy to change titles to accommodate new perfumes. See figs. 9.31, 9.50 and 9.256.

14.18   THE ESCUTCHEON

This is the classic imposing wine label shape conveying a feeling of importance such as is conveyed by an armorial bearing. For some examples see CRÈME DE PORTUGALE, HAMMAM BOUQUET, HUILE D'ANIS DES INDIES, HUILE D'ANIS ROUGE, PEROXIDE and MILK PUNCH. Some labels have "ears" attached which have chains affixed to them to enable labels to hang properly on the bottle. Examples are CLOVES and EAU DE COLOGNE. See fig. 9.60.

14.19   THE COUNTRYSIDE

This is an individualistic design produced by Wakeley and Wheeler in 1956 on country themes for use in a house in the countryside. See fig. 9.68.

14.20   THE HAT

Eau de Nil, an unmarked label in gilt-bronze, follows the shape of a floppy hat. Rawlings and Summers' silver-gilt CURACOA of 1849 is in similar style. See fig. 9.89.

14.21   THE ARCHITECTURAL

This is a well known sauce label design, no stranger to Rawlings and Summers who used it in 1836 for LAVENDER WATER and ROSE WATER. See fig. 9.219c.

14.22   THE VASE OF FLOWERS

This French design sums up and captures the purpose of VERVEINE. It was also used for ANISETTE. See fig. 9.13.

14.23   THE HUNT

An example for Sweet is illustrated. Hunting scenes were popular. The Queen Mother at Clarence House had hunting scene menu card holders in the Dining Room, a gift from America. See fig. 9.229b.

14.24   THE SINGLE CUPID

The cupid in this design peers above or over a festooned swag. This is a Bateman design. See fig. 9.54.

14.25   THE TWO CUPIDS

Joseph Willmore used his two cupid wine label for EAU D'OR in 1814. There is also a German version of this design for COLOGNE which was displayed at the Circle's AGM in 1978. See fig. 9.92.

14.26   THE BANNER

This long rectangular shaped flag design was used for a silver label inscribed for Gomme, a pick-me-up made of gum syrup, in lower case lettering. See fig. 9.102.

14.27   THE SCROLL

This design was employed upon a Royal Occasion for CHERRY. See fig. 9.54.

14.28   THE BATTERSEA STYLE ENAMEL

This style is evidenced by the quality of the decoration. Much has been written about it. See figs. 9.77 and 9.179a.

14.29   THE ARMORIAL

Some owners liked to personalise their labels which adds interest. A crested banner or scroll bearing the title LEMON shows an heraldic beast above the mantling. See fig. 9.161f.

14.30   THE ARTS AND CRAFTS

Whilst the design and shape of a label readily identifies a boudoir label, its boudoir purpose, probably enhaced by the Arts and Crafts Movement and the influence of Guilds of Handicrafts, is confirmed by the use of enamel lodged in the silver especially in the title of the label and sometimes by the use of paste or semi-precious stones.

Fig 15.1

CHAPTER 15

RECIPE BOOKS

A selection of these authoritative books is shown in Fig. 15.1

A bibliography of recipe books referred to in this book, such as Dr. Scott's House Book, the title page of which is illustrated in fig. 15.2, is as follows:

15.1   1737. Eliza Smith. The Compleat Housewife or Accomplished Gentlewoman's Companion, including above 300 Family Receipts for Medicines (8th Edition).

15.2   1774. Hannah Glasse. The Art of Cookery made plain and easy, including 150 new receipts (6th Edition).

15.3   1785. William Henderson. The Housekeeper's Instructor or Universal Family Cook (7th Edition).

15.4   1799. Elizabeth Raffald. The Experienced English Housekeeper (12th Edition).

15.5   1804. John Farley. The London Art of Cookery and Housekeeper's Complete Assistant (10th Edition).

15.6   1824. Elisabeth Insull. The Female Instructor or Young Woman's Companion and Guide to Domestic Happiness (2nd Edition).

15.7   1826. Doctor William Scott. The House Book or Family Chronicle of Useful Knowledge and Cottage Physician (1st Edition).

15.8   1844. Lady Charlotte Bury. The Lady's Own Cookery Book and New Dinner Table Directory (3rd Edition).

15.9   1855, Frederick Bishop. The Wife's Own Book of Cookery (1st Edition).

15.10 1857. Robert Kemp Philp. The Housewife's Reason Why affording to the Manager of Household Affairs intelligible reasons for the various duties she has to superintend or perform (1st Edition).

15.11 1864. Cre – fydd. The Young Housewife's Daily Assistant on all matters relating to Cookery and Housekeeping (2nd Edition).

15.12 1870. Isabella Beeton. The Englishwoman's Cookery Book taken from her Book of Household Management (New Edition).

15.13 1910. Isabella Beeton. On The Art of Cookery and all about Cookery and Household Work (New Edition).

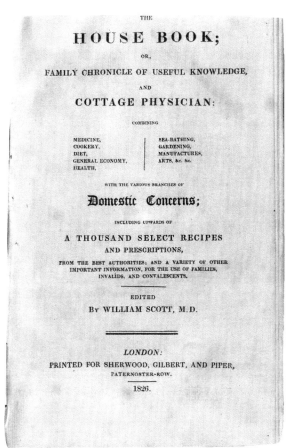

Fig 15.2

CHAPTER 16

ILLUSTRATIONS

A list of the illustrations used in this book,
with figure references given, is as follows:

ILLUSTRATIONS

CHAPTER 17

INDEX

Each of the foregoing Chapters has paragraph numbers to which reference is made in this index. Illustrations are identified by reference to figure numbers and are separately indexed in Chapter Sixteen.

INDEX

INDEX

INDEX

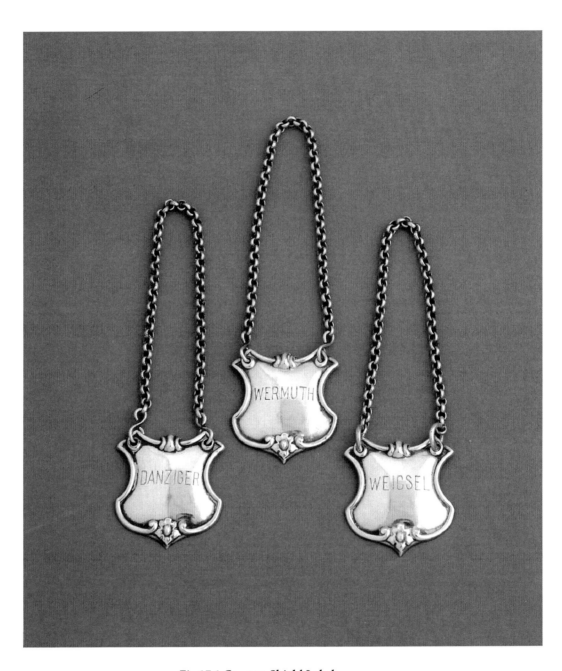

Fig 17.1 German Shield Labels